What a difference in VALUE a few years make!

NOW . . for only $89.95*

RCA Victor

ELECTRIC TUNING!

RCA Victor Electric Tuning Model 87K1 . . . In addition to Electric Tuning this beautiful new instrument offers the famous Magic Eye and RCA Metal Tubes. Straight-Line Dial, 12" dynamic speaker, phonograph connection. American-foreign reception, police, aviation and amateur calls. $89.95*.

Compare the value! The advertisement on the opposite page tells about the Aeriola Grand selling for $409.50. That was in 1922.

And now—look at present-day radio value! A beautiful 7-tube RCA Victor Radio with Electric Tuning—for only $89.95*! In all radio history few developments have captured the public fancy as quickly and completely as has RCA Victor Electric Tuning! Just imagine the thrill of getting any one of your six favorite stations with the simple push of a button! That's all there is to it! Just push a button—there's your station, and it comes in tuned "right on the nose."

Visit your nearest RCA Victor dealer today or tomorrow. See this amazing instrument that can be yours for so little. Notice its many other fine features. Every one of them is *proof* that RCA Victor offers great radio value—and that *this* set is one of the greatest values of all time!

• • •

When buying radio tubes, say "RCA". First in Metal —Foremost in Glass—Finest in Tone.

RCA presents the Magic Key every Sunday, 2 to 3 P.M., E. S. T., on NBC Blue Network.

**Price f. o. b. Camden, New Jersey, subject to change without notice.*

RCA MANUFACTURING COMPANY, INC. CAMDEN, NEW JERSEY

A SERVICE OF THE RADIO CORPORATION OF AMERICA

THIS RADIO CORPORATION OF AMERICA ADVERTISEMENT WAS PUBLISHED IN 1914

THIS ADVERTISEMENT APPEARED IN 1914

THIS RADIO CORPORATION OF AMERICA ADVERTISEMENT WAS PUBLISHED IN 1926

Announcing the National Broadcasting Company, Inc.

National radio broadcasting with better programs permanently assured by this important action of the Radio Corporation of America in the interest of the listening public

THE RADIO CORPORATION OF AMERICA is the largest distributor of radio receiving sets in the world. It handles the entire output in this field of the Westinghouse and General Electric factories.

It does not say this boastfully. It does not say it with apology. It says it for the purpose of making clear the fact that it is more largely interested, more selfishly interested, if you please, in the best possible broadcasting in the United States than anyone else.

Radio for 26,000,000 Homes

The market for receiving sets in the future will be determined largely by the quantity and quality of the programs broadcast.

We say quantity because they must be diversified enough so that some of them will appeal to all possible listeners.

We say quality because each program must be the best of its kind. If that ideal were to be reached, no home in the United States could afford to be without a radio receiving set.

Today the best available statistics indicate that 5,000,000 homes are equipped, and 21,000,000 homes remain to be supplied.

Radio receiving sets of the best reproductive quality should be made available for all, and we hope to make them cheap enough so that all may buy.

The day has gone by when the radio receiving set is a plaything. It must now be an instrument of service.

WEAF Purchased for $1,000,000

The Radio Corporation of America, therefore, is interested, just as the public is, in having the most adequate programs broadcast. It is interested, as the public is, in having them comprehensive and free from discrimination.

Any use of radio transmission which causes the public to feel that the quality of the programs is not the highest, that the use of radio is not the broadest and best use in the public interest, that it is used for political advantage or selfish power, will be detrimental to the public interest in radio, and therefore to the Radio Corporation of America.

To insure, therefore, the development of this great service, the Radio Corporation of America has purchased for one million dollars station WEAF from the American Telephone and Telegraph Company, that company having decided to retire from the broadcasting business.

The Radio Corporation of America will assume active control of that station on November 15.

National Broadcasting Company Organized

The Radio Corporation of America has decided to incorporate that station, which has achieved such a deservedly high reputation for the quality and character of its programs, under the name of the National Broadcasting Company, Inc.

The Purpose of the New Company

The purpose of that company will be to provide the best program available for broadcasting in the United States.

The National Broadcasting Company will not only broadcast these programs through station WEAF, but it will make them available to other broadcasting stations throughout the country so far as it may be practicable to do so, and they may desire to take them.

It is hoped that arrangements may be made so that every event of national importance may be broadcast widely throughout the United States.

No Monopoly of the Air

The Radio Corporation of America is not in any sense seeking a monopoly of the air. That would be a liability rather than an asset. It is seeking, however, to provide machinery which will insure a national distribution of national programs, and a wider distribution of programs of the highest quality.

If others will engage in this business the Radio Corporation of America will welcome their action, whether it be cooperative or competitive.

If other radio manufacturing companies, competitors of the Radio Corporation of America, wish to use the facilities of the National Broadcasting Company for the purpose of making known to the public their receiving sets, they may do so on the same terms as accorded to other clients.

The necessity of providing adequate broadcasting is apparent. The problem of finding the best means of doing it is yet experimental. The Radio Corporation of America is making this experiment in the interest of the art and the furtherance of the industry.

A Public Advisory Council

In order that the National Broadcasting Company may be advised as to the best type of program, that discrimination may be avoided, that the public may be assured that the broadcasting is being done in the fairest and best way, always allowing for human frailties and human performance, it has created an Advisory Council, composed of twelve members, to be chosen as representative of various shades of public opinion, which will from time to time give it the benefit of their judgment and suggestion. The members of this Council will be announced as soon as their acceptance shall have been obtained.

M. H. Aylesworth to be President

The President of the new National Broadcasting Company will be M. H. Aylesworth, for many years Managing Director of the National Electric Light Association. He will perform the executive and administrative duties of the corporation.

Mr. Aylesworth, while not hitherto identified with the radio industry or broadcasting, has had public experience as Chairman of the Colorado Public Utilities Commission, and, through his work with the association which represents the electrical industry, has a broad understanding of the technical problems which measure the pace of broadcasting.

One of his major responsibilities will be to see that the operations of the National Broadcasting Company reflect enlightened public opinion, which expresses itself so promptly the morning after any error of taste or judgment or departure from fair play.

We have no hesitation in recommending the National Broadcasting Company to the people of the United States.

It will need the help of all listeners. It will make mistakes. If the public will make known its views to the officials of the company from time to time, we are confident that the new broadcasting company will be an instrument of great public service.

RADIO CORPORATION OF AMERICA

OWEN D. YOUNG, Chairman of the Board

JAMES G. HARBORD, President

THIS ADVERTISEMENT APPEARED IN 1926

In every sense of the word

NBC *is*

"Broadcasting Headquarters"

Take PROGRAMS . . . NBC's Red and Blue Networks sent out over 54,000 programs during 1937. 35 hours a day of the world's most popular programs, available to 24,500,000 radio families.

Take STATIONS . . . Completing its eleventh year, there are 142 affiliated stations broadcasting Blue and Red Network programs. Ten NBC associated short-wave stations make NBC the leader in international broadcasts.

Take PROGRAM BUILDING . . . For every one of these eleven years, NBC has set the style in programs. These styles, numbering many "Famous Firsts" in radio, have since become radio patterns.

Take TRANSCRIPTIONS . . . The NBC Recorded Program Service offers complete facilities for the creation, casting, production and recording of programs for National Spot and Local Advertising. NBC Thesaurus is used by no less than *220 station subscribers!*

Whether you Write it . . . Wire it . . . Phone it . . . Cable or Radio it

"RADIO CITY
NEW YORK"
is the World's best known Radio Address

Take SPOT and LOCAL ADVERTISING . . . NBC's Managed Stations are, everywhere, "Tops in Spot!" Super-power transmitters on clear channels cover major markets, making an economical lineup for spot advertisers who want to cover broad territories with a few stations. *Fifteen fine stations in ten key markets.*

Take ARTISTS SERVICE . . . NBC Artists Service is the largest talent sales organization in the world. It offers both personal management and valuable guidance to artists. To radio advertisers and their agencies, it offers not only specific talent to answer advertisers' sales problems, but sound program ideas as well, and complete casts.

. . . *These advantages, plus NBC's known primary concern that "the listener be served", first, have caused NBC to be known, wherever radio is mentioned, as* "BROADCASTING HEADQUARTERS"

NATIONAL BROADCASTING COMPANY

A RADIO CORPORATION OF AMERICA SERVICE

THIS RADIO CORPORATION OF AMERICA ADVERTISEMENT WAS PUBLISHED IN 1920

WORLD WIDE WIRELESS

IN addition to the complete line of Radio Transmitting and Receiving sets sold by the RADIO CORPORATION OF AMERICA, for Ship to Ship and Ship to Shore Communication, the Corporation is in a position to quote upon:

PORTABLE MULE PACK AND CART RADIO SETS.

SPECIAL TYPES OF SPARK TRANSMITTERS FROM 1 TO 20 KILOWATTS, FOR MOBILE AND FIXED STATIONS FOR COMMERCIAL AND GOVERNMENT USE.

MILITARY TRACTOR SETS OF THE VACUUM-TUBE TYPE.

SPECIAL CONTINUOUS WAVE TRANSMITTERS FOR SUBMARINES, DESTROYERS, SEAPLANES AND FOR NAVAL BASES.

SPECIAL INTER-FLEET RADIO TELEPHONE SYSTEMS FOR BATTLESHIPS AND OTHER NAVAL CRAFT.

The Corporation has in development a new series of Continuous Wave Tube Transmitters for the various fields of commercial usage. This apparatus bears the stamp of The General Electric Co., (U.S.A.) one of the world's largest electrical manufacturers.

TRANS-OCEANIC RADIO

Two hundred kilowatt radio frequency alternators are now in daily use in the Corporation's high-power trans-oceanic stations. The Corporation is in a position to bid on similar equipment for international communication.

SHIP-WIRELESS may be bought outright, may be leased, and may be bought or leased in addition to our service of inspection, maintenance and operation. We provide licensed, skillful and trustworthy operators. Also our service and inspection depots are located at all important seaports of the world.

Our Nearest Office Will Give You All Details and Information

RADIO CORPORATION OF AMERICA

EDWARD J. NALLY, President

WOOLWORTH BUILDING NEW YORK CITY

BRANCH OFFICES IN THE U. S. A.

BOSTON
NEW ORLEANS
SEATTLE

PHILADELPHIA
GALVESTON
PORTLAND

BALTIMORE
PORT ARTHUR
SAN FRANCISCO

NORFOLK
CLEVELAND
SAN PEDRO

SAVANNAH
CHICAGO
HONOLULU

THIS ADVERTISEMENT APPEARED IN 1920

THIS RADIO CORPORATION OF AMERICA ADVERTISEMENT WAS PUBLISHED IN 1928

Candidates for the Promised Land

"The Promised Land" is the name professional operators have given to the New York Radio Central Station, on Long Island. When completed, this station will be the largest and most powerful radio station in the world. It will be equipped to work simultaneously with five other nations in widely separated and distant parts of the world, and will be epoch-making in the field of international communication.

Enrollments are coming in by every mail. Why aren't you one of the wideawake wireless men who have seen the new and greater opportunity opened to them by this fascinating field? NOW—not some later day—is the time to act!

A position at this station is the height of every operator's ambition, for it means unlimited opportunity to succeed and progress to higher, more responsible and better paying positions in the radio industry. So far as opportunity goes the successful future of these men is assured.

How about you?

This new course of radio training, which has been developed for the benefit of those who cannot attend the Institute personally, is the same course used at the Institute. It includes everything from basic principles of electricity and magnetism, to actual operation of commercial radio equipment. It also includes the same textbooks used in the Institute classes, as well as a buzzer set of greatly improved design, with a variable automatic transmitter, for code practice.

The graduates of the Radio Institute of America enjoy a great and exclusive advantage in the close connection existing between the Institute and the Radio Corporation of America, world's largest radio manufacturing and commercial radio company.

Prominent executives in the radio field are former students of the Institute. The Radio Corporation employs thousands of men, in its executive departments, on ships and at shore stations and in factories and laboratories. A large percentage of these men are graduates of the Institute.

The Radio Institute of America has been an established and successful institution for over fifteen years. The year round average attendance in its classrooms is now 298 students per month. It has trained over 6,000 men, 95% of whom have successfully engaged in this new branch of science and industry.

You, too, can be successful in this new field if you properly train yourself. Radio offers an unlimited opportunity for future advancement—why not take advantage of it.

Write for our booklet and further details—*now*.

Radio Institute *of* America

(formerly Marconi Institute)

326 Broadway, New York

THIS ADVERTISEMENT APPEARED IN 1928

at last— the *portable* super-heterodyne

PERFECTED

by McMurdo Silver. It surpasses your fondest hopes.

AUTHORITIES

on Radio have declared it to be an "ELECTRICAL MASTERPIECE." It secures results you never hoped to obtain outside of a Laboratory, and is so simple that you can construct it with a pair of Pliers, a Screw Driver and a Soldering Iron.

THE PARTS

recommended by Mr. Silver include a drilled and engraved Panel, and everything necessary to build the set. Price $58.00 Parcel Post prepaid, East of Rockies. (Accessories not included: Tubes, Batteries, Cabinet, Loop, Loud Speaker).

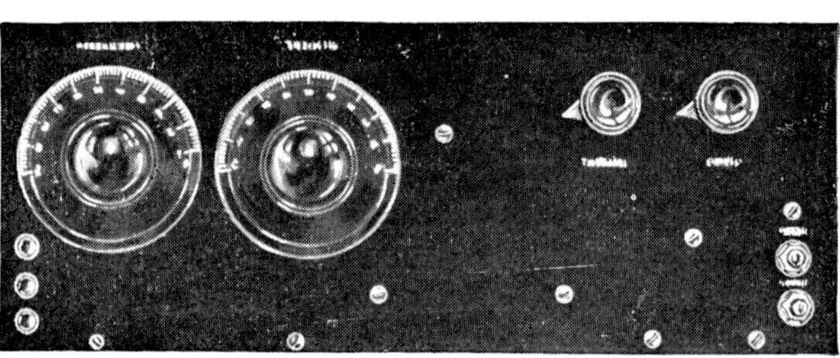

SILVER SPECIALS

Oscillator Coupler, No. 101	$2.50
30 KC Tuned Output Transformer, No. 201	3.50
.0005 Low Loss Condenser, No. 301	4.50
50 KC R.F. Transformer Unit, No. 401	4.00
5-Gang 199-Socket, No. 501	3.00

Write for Descriptive Circulars

All S-M Products are backed by Silver-Marshall's Unconditional Guarantee of Satisfaction or Your Money Back.

EASTERN DISTRIBUTOR
Twentieth Century Radio Corp., 102 Flatbush Avenue
Brooklyn, N. Y.

Silver-Marshall, inc.

105 S. Wabash Avenue, Chicago, Ill.

THIS ADVERTISEMENT APPEARED IN 1924

Still "Years Ahead"

RADIO BY McMURDO SILVER

STARTLING radio engineering achievements have characterized the work of McMurdo Silver for a quarter of a century. Year after year his radio receivers have won the acclaim of technical schools, engineering laboratories, radio distance receiving champions, leaders of scientific expeditions, and internationally famous musicians . . . have made widely accessible results previously considered unobtainable outside of a laboratory. This leadership in radio engineering is strikingly demonstrated by current models. It is no exaggeration to say that no real conception of what radio offers *today* is possible until you have tested the 1938 masterpieces of this master engineer. Here is radio which, with uncanny "reach" goes to the earth's four corners to snare unique and unusual programs ... music, sports, news, ships' messages. Thirty-six distinctive technical features are responsible for the "Years Ahead" performance of tomorrow's radio by McMurdo Silver. Send for details.

The 14-15—*Super-sharp communication receiver of 14 tubes (15 with larger speaker) with tuned and stabilized regenerative r. f. amplification on all wave bands.*

The 15-17—*Advanced 15 tube super-heterodyne circuit delivering functional efficiency of 17 tubes. Every coil and circuit individually shielded and chassis chrome plated throughout.*

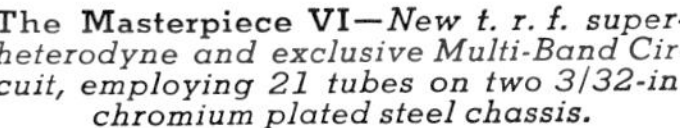

The Masterpiece VI—*New t. r. f. super-heterodyne and exclusive Multi-Band Circuit, employing 21 tubes on two 3/32-in. chromium plated steel chassis.*

Type 1B Pre-Amplifier Mixer—

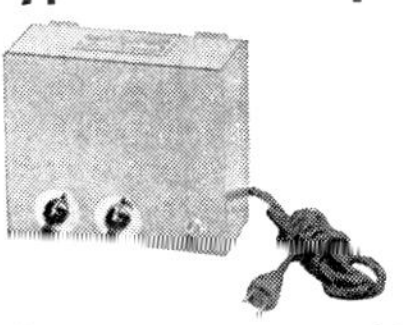

This pre-amplifier and electronic-mixer (self powered, universal a.c.—d.c.) features two channels into the two triode sections of a 6F8, the two capable of being mixed and faded together by individual gain controls. The first with a 6J7 pentode yields 50 db. voltage gain with an input of 2 megohms permitting use of crystal or other low-level microphone plugged in directly through standard shielded jack and plug. The second channel serves for high-level microphone and phono-pick-up or radio receiver input and gives a voltage gain of 20 db. Operation is from any 105 to 125 volt, 25 to 60 cycle a.c. or d.c. power line. Hum is unmeasurable and the 1B may be used to drive following power amplifiers with no trouble on this score, with 20 to 50 foot separation through single conductor shielded microphone cable. Substantial steel shielding case is 7" long, 6" high and 3-1/2" deep. Shipped complete with 1 each tested Raytheon 6J7, 6F8 and 6J5 it will turn any good receiver into a complete P. A. System.

New and Sensitive Dynamic Microphone

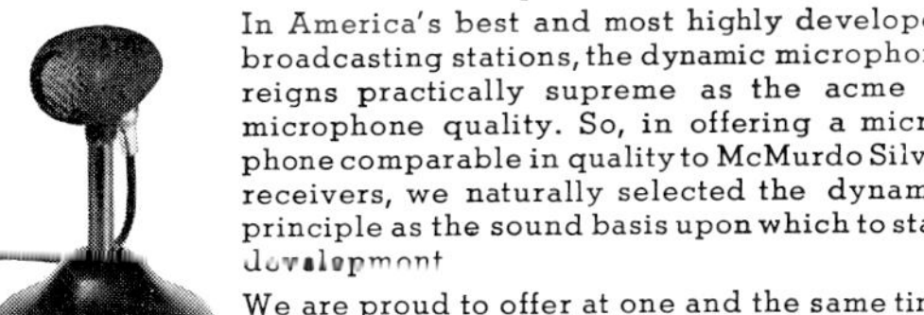

In America's best and most highly developed broadcasting stations, the dynamic microphone reigns practically supreme as the acme of microphone quality. So, in offering a microphone comparable in quality to McMurdo Silver receivers, we naturally selected the dynamic principle as the sound basis upon which to start development.

We are proud to offer at one and the same time not only a microphone directly comparable in quality to the costly units used in the finest broadcasts, but to provide through recent research greatly increased sensitivity, a combination of directional or non-directional characteristics in one single unit and to be able to establish what is an extraordinarily low price for such a remarkable instrument.

The new McMurdo Silver 4A Dynamic Microphone is shown mounted upon its 7" high, no-tip-over steel desk stand, together with the shielded plug which terminates at the microphone, the 25-foot rubber-covered shielded microphone cable with which it is equipped. Both stand and cable plug unscrew for convenience. The stand thread is standard, so that the microphone itself may be quickly transferred to different stands as may be desired.

This superb microphone is only 3-13/16" long and 2-1/4" in diameter, and weighing but 1-1/2 lbs. has a frequency characteristic flat to 3 db. from 50 to 10,000 cycles—the full range of the best broadcast stations. Its "level" is unusually high, being only—52 db. and its output impedance is 50,000 ohms and materially reduces a.c. hum pick-up by the microphone or its cable and permits quite long microphone-to-amplifier cables—even up to a couple of hundred feet when necessary.

New Selective-Directional Antenna

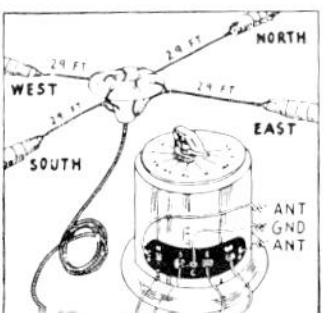

The SELECTIVE-DIRECTIONAL 9 is a positive means of boosting weak signal strength and reducing noise and interference. It provides an extraordinary amount of local noise reduction—depended wholly upon how far away from and high above local electrical noise sources one can erect its "X" flat-tops.

The SELECTIVE-DIRECTIONAL 9 gives nine different choices of directional doublet, single wire and V-aerials. One may select at the turn of the knob an East-West or North-South doublet, an East, West, or South single wire "L" antenna, or a South-East, South-West, North-West or North-East "V" aerial.

All are coupled to the receiver through a noise rejecting transmission line and perfected automatically self-selecting long and short wave antenna and set couplers.

All soldered and ready to put up, erection requires only two crossing 60-foot spans in approximate "X" form on house top, open lot, or anywhere such spans can be conveniently found.

Kit includes 4 antenna and insulators, 4 stand-off insulators, lead-in insulator tube and 50-ft. of weather-proof lead-in cable.

McMURDO SILVER

2900-G SOUTH MICHIGAN AVENUE, CHICAGO, ILLINOIS

1910

Wireless Equipment Revolutionised.

Over a year ago we began applying our Condensers for Wireless Work, and we have now supplied a large number for voltages up to 30,000. They have been subjected to very severe tests, and we are receiving repeat orders from British, Colonial and Foreign Governments, and from other large users in all parts of the world.

We claim, and tests have proved, that Dubilier Condensers are more efficient than any others; that they occupy not more than **one-tenth of the space** and are about **one-tenth the weight** of other high-tension condensers; that they cost no more than other good condensers; that they cannot be surpassed for portability; that they are practically **unbreakable**; that there are no losses due to brush discharge or leakage; that hysteresis is reduced to a minimum; that desired capacities are supplied true to within 1 per cent.; that the efficiency of wireless stations is increased by the installation of Dubilier Condensers, and therefore

The Sending Radius is Greatly Increased.

Working Voltage ... 30,000	Working Voltage ... 15,000	Working Voltage ... 20,000	Capacity ... ·25 M.F.
Breakdown " ... 75,000	Breakdown " ... 42,000	Breakdown " ... 50,000	Breakdown Voltage 3,000
Capacity ... ·006 M.F.	Capacity ... ·015 M.F.	Capacity ... ·003 M.F.	Dimensions, 10·5×8×3 C.M.
Dimensions, 18×15·5×8·5 C.M.	Dimensions, 16×13·5×9 C.M.	Dimensions, 13·5×8·5×8·5 C.M.	

***SEND YOUR ENQUIRIES.* All Condensers Guarante**

and are returnable if they do not give entir

THE DUBILIER ELECT

96 9

THIS ADVERTISEMENT APPEARED IN 1910

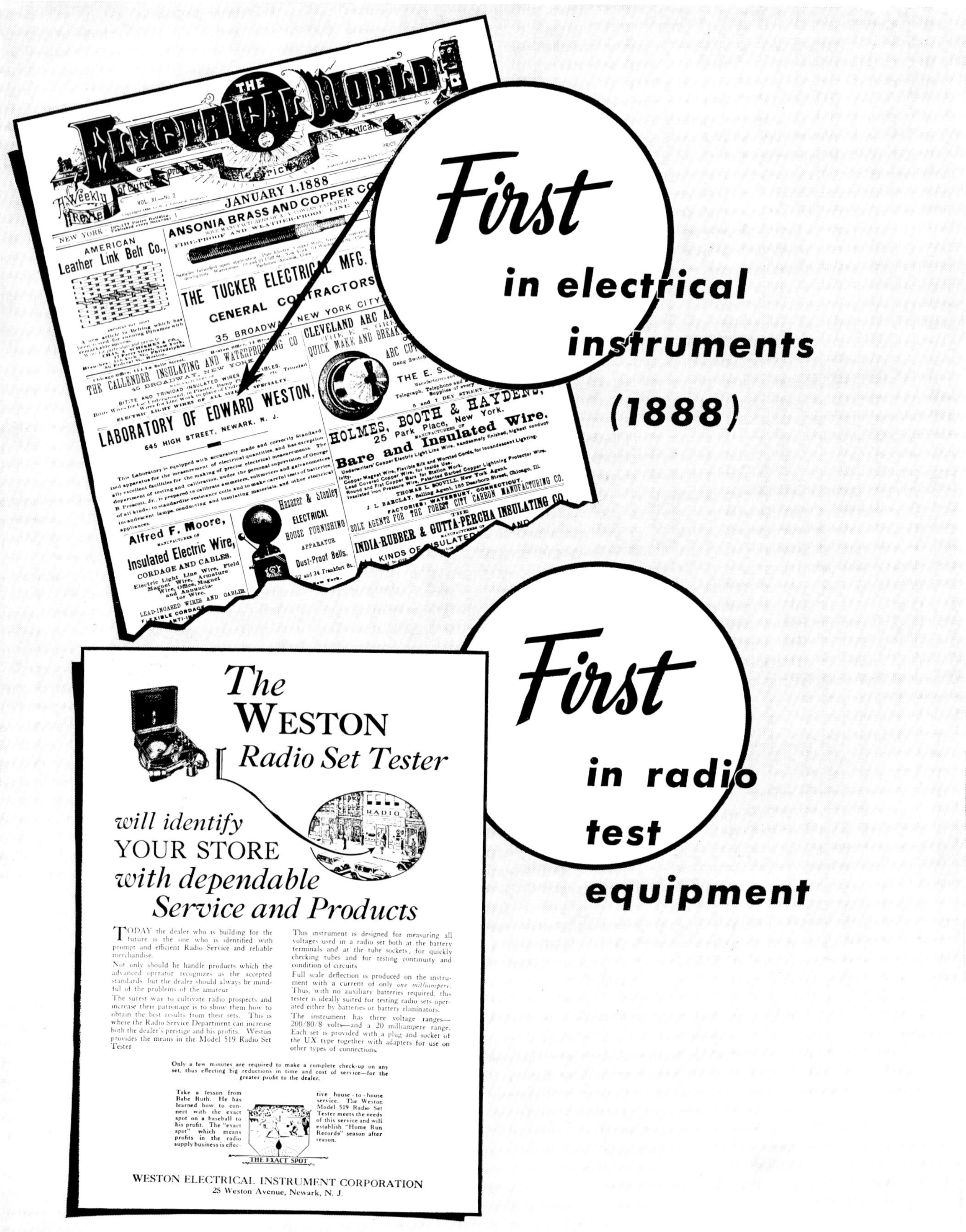

THESE ADVERTISEMENTS APPEARED IN 1888 AND 1925

No matter what type of test equipment is needed, radio servicemen know that when they buy WESTON they get the sound engineering, <u>the added dependability,</u> which comes from a half-century experience in building better instruments. Send for literature on the complete line. Weston Electrical Instrument Corp., 599 Frelinghuysen Ave., Newark, N. J.

WESTON *Radio Instruments*

Do Amateurs Realize the Wireless Opportunities that Await Them?

How the President of the National Radio Institute Answered this Question When It Was Put Up to Him. What Would You Have Said? Is the World's Fastest-Growing Field Actually Going to Slip Away From Those Best Able to Cash In Big On It? These Are Questions Which Will Interest Every Radio Amateur.

Mr. James E. Smith, President, National Radio Institute.

THAT was one of the questions recently put up to me by a well-known authority visiting Washington. "In your opinion," he said, "do amateurs realize the wireless opportunities that await them?" For a moment I was stumped! Then I replied, "Yes, with just one 'but'. I think that amateurs are well aware of the tremendous expansion of wireless that is daily going on. They realize that it is sweeping the world like wildfire. BUT I do not think that they realize what this means to *them*—they do not realize that they can easily get the 'plums' that the field offers. They 'have the jump' on everyone else, and they should realize now that 'the fastest-growing field in the world' besides being a fascinating hobby is a wonderful, opportunity-filled field offering splendid present advantages—and growing so rapidly that the future is beyond estimation!"

I wonder if many amateurs have ever considered the fact that what is to them a fascinating hobby is also a fascinating profession, filled with big opportunities that they can easily share whenever they are ready to do so. It's only a short step for them now to a splendid field that they can put their hearts into—and offering a bigger future than old businesses which are overcrowded.

Big Opportunities Are Knocking—Are Some of Us Saying "Please Go 'Way and Let Me Sleep?"

After the caller who started me thinking about this matter had left, I jotted down on my pad some of the items which I had recently noted regarding wireless expansion. On land and on sea big opportunities are opening, and even greater uses for wireless are being found every day. No doubt you too have read these items, but I am going to have them printed here because I want to impress upon you what this tremendous expansion can mean to you.

When I read every day how wireless expansion is sweeping over the world I often say to myself, "Big opportunities are knocking—I wonder if amateurs realize that they can cash in big on this growing field. While opportunities knock, I wonder if some aren't saying, 'Please go 'way and let me sleep'." Of course, they aren't sleeping by any means, but I want *all* of them to know just how easy it is to fully qualify for a field which is undeniably filled with greater advantages than most others in the world today.

Easy to Qualify In Spare Time—at Home

I want to tell you—without obligation to yourself in any way—more about wireless opportunities and how you can take advantage of them. I would like to tell you about our Institute. This National Radio Institute was the original and is today the oldest and largest school in America teaching wireless by mail. The government allows our graduates five to ten points credit when taking First Grade Government License examinations. We have graduates in almost every part of the world who have quickly qualified through the special method through which we make Wireless amazingly easy for anyone to learn completely at home in spare time.

These are some of the main points about this Institute and I am sorry I haven't room to tell you all of them. I should like to tell you more about our wonderful new methods of teaching, about our remarkable new invention, the "Natrometer," which each student gets and which almost cuts in half the time necessary to learn Wireless thoroughly. Then too I'd like to tell you about our Diploma, our Relay League, Employment Service, and about our special easy-payment plan. But there is not enough room here to tell you about all these things so I am going to ask you to write me for a new interesting booklet we have gotten up.

Write Me For Booklet

A little coupon is being put here so that you can save yourself trouble in sending for this illustrated booklet, "Wireless, the Opportunity of Today." By mailing this coupon you will not be obligating yourself in any way and no solicitor will call upon you. But the coupon *will* bring you some mighty interesting facts about Wireless Opportunities and about how you can quickly and easily qualify for them—at home and in your spare time.

Won't you mail this little coupon at once? Whether you are a junior Radio Amateur and want to learn all about Wireless or whether you are anxious to fully qualify so as to enter the wireless profession now in one of the fine opportunities open on land or on sea—write me for this booklet. All that I ask is that you write as soon as possible. And—since there is no obligation—why not write me today!

P. S.—By the way, we are making a special short-time offer, for a strictly limited time, in which we are giving all new students our complete new course in Wireless Telephony FREE. Mail the coupon direct to me, today, and let me tell you about it by return mail.

Mr. James E. Smith, President, The National Radio Institute, Dept. 16, Washington, D. C.

MAIL THIS COUPON TODAY

Mr. James E. Smith, President,
National Radio Institute,
Dept. 16, Washington, D. C.

Send me your FREE book, "Wireless, the Opportunity of Today." Tell me about your Institute and your special short-time offer.

Name ....

Age .... Occupation ....

Address ....

City .... State ....

What I Jotted Down

Here are the items I jotted down on my pad, showing how Wireless is growing by leaps and bounds all over the world. Let me tell you what this world-wide sweep of wireless expansion means to you and to your future.

A $20,000,000 American corporation has been formed to establish wireless stations in every part of the globe.

The U. S. Merchant Marine operates thousands of vessels. Wireless is now a necessity on ships.

The Chicago Tribune now receives foreign news by wireless. Other papers are calling upon Wireless too.

Huge wireless stations are springing up all over the world. Saint Assise, France; Bordeaux, Ville Juif, and Lyons, France; Peking, China; Geneva, Switzerland; Shanghai, China; Fiji Islands; Warsaw, Poland—and these are but a few.

Many railroads are calling upon wireless to dispatch trains and carry on communication. The Lackawanna, The Louisville & Nashville, The Canadian-Pacific, The Nashville, Chattanooga & St. Louis, are some of them. ——New York, Cleveland, Chicago and Detroit are connected by an inter-city wireless service.

Criminals are being intercepted by wireless through the Police Department of New York, Dallas, Chicago, and other cities.

Brokers, Bankers, Merchants, Manufacturers and other business concerns are calling upon wireless. John Wanamaker, Goodyear Rubber Co., Standard Oil Co., New York Stock Exchange, are only a few.

Farmers are getting Market and Weather reports daily by wireless in all sections of the country.

New wireless stations are springing up in every part of America. Belfast, Maine; Cape May, N. J.; East Pittsburgh, Pa.; San Francisco, Cal.; Helena, Montana; Seattle, Washington; Mobile, Alabama—these are but a few.

The Aerial Mail Service of the Post Office Department already has 12 radio stations in operation.

The Japanese are constructing a powerful station in the Orient.

A big new wireless service is being established between England and France.

The Federal Telegraph Co. is establishing a complete chain of stations on the Pacific Coast.

Messages are sent from the Philippine Islands to Washington (10,000 miles) in 3 minutes.

Daily wireless service between the United States and Japan is in full operation.——St. Johns, New Foundland, is operating a large service.

Danzig, in Europe, is carrying on large wireless operations.

Three tremendous stations are operating on Long Island at Easthampton, Port Jefferson, and East Moriches.

South America is planning to establish a chain of stations at Rio de Janeiro, Asuncion, Buenos Aires and Montevideo.

One single American concern offers wireless communication between the United States and France, England, Germany, Norway, Denmark, Sweden, Finland, Poland, Honolulu and Japan.

And these are only a few of the examples showing how Wireless expansion is spreading over the whole earth. It brings you amazing opportunities and you can now easily grasp them.

THIS ADVERTISEMENT APPEARED IN 1921

SAY FELLOWS
GET INTO RADIO-TELEVISION AND TALKING PICTURES

Let me tell you how I can quickly train you, NOT by book study, but by actual shop training on real Radio, Television and Talking Picture equipment in 12 WEEKS in the great shops of COYNE in Chicago.

Here at Coyne you don't need advanced education or experience and many of my students Earn while learning. After graduation I give them lifetime employment service. Here at Coyne too you get individual instruction and you can start anytime.

Radio offers jobs as Designer, inspector and tester, salesman, and in installation work, operator of a broadcasting station, wireless operator on a ship, with Talking Picture theatres — with Television Laboratories and studios. Television alone will soon be calling for thousands of trained men.

Come to Coyne here in Chicago and prepare for one of these jobs the quick and practical way — BY ACTUAL SHOP WORK.

It's a shame for any fellow to go thru life as an untrained man working at small pay, and never even sure of a steady job, and when he does work working at any old price they want to pay him.

You can avoid this. You can be a trained man and have a real future. Mail the coupon today and I'll send you my big Free book and tell you how you can be a success just as hundreds of my graduates are. H C Lewis

H. C. LEWIS, President
Radio Division, COYNE ELECTRICAL SCHOOL
500 S. Paulina St., Dept. 38-7H, Chicago, Ill.
Send me your Big Free Radio and Television Book, and tell me how I too can make a success in Radio.

Name ....

Address ....

City .... State ....

REPRODUCED FROM AN OLD ADVERTISEMENT

MANY OPPORTUNITIES FOR THE COYNE TRAINED RADIO MAN

Are You Ready for a Better Job-More Pay?

Don't be an untrained man. Let me show you how to get your start in Radio — a fast growing, live money-making industry.

Prepare for jobs as Assembler, Inspector and Tester—Radio Sales or Service and Installation Work—Broadcasting Station Operator—Wireless Operator on a Ship or Airplane or Sound Work — HUNDREDS OF OPPORTUNITIES for a real future in radio!

12 Weeks of Shop Training

We don't teach by book study. We train you on a great outlay of Radio, Television and Sound equipment—on scores of modern Radio Receivers, actual Broadcasting equipment, Television apparatus and Sound Reproduction equipment, Code and Telegraph equipment, etc. You don't need advanced education or previous experience. We give you — RIGHT HERE IN THE COYNE SHOPS — the actual practice and experience you'll need for your start in this great field. And because we cut out all useless theory and only give that which is necessary you get a practical training in 12 weeks. Mail coupon for all facts about my school and training methods.

TELEVISION *and* PUBLIC ADDRESS

Television is sure to come as a commercial industry. Rapid progress is now being made in developing this new field. It will offer opportunities to the man who is trained in Radio. Here at Coyne you learn Television principles, and work on actual Television equipment. Public Address Systems offer opportunities to the Trained Radio Man. Here is a great new Radio field which is rapidly expanding. Prepare NOW for these wonderful opportunities! Learn Radio Sound Work at COYNE on actual **Sound Reproduction** equipment. **Not a home study course.**

SEND FOR DETAILS OF MY

"PAY AFTER GRADUATION" PLAN

Mail the Coupon below and I'll tell you about my "Pay After Graduation" Plan **which has enabled hundreds of others to get Coyne training** with **very little money.** On this plan you can get your training first, then **take 18 months** to complete your small monthly tuition payments starting **5 months** after you begin training. **Not a home study course.**

Mail the coupon for all details of this **"Tuition Payment Plan."**

PRACTICAL WORK
at COYNE in Chicago

ACTUAL, PRACTICAL WORK. You build and service radio sets. You get training on real Broadcasting equipment. You construct Television Receiving Sets and **actually transmit your own Television images over our Television equipment.** You work on real Sound equipment. You learn Wireless Operating on Actual Code Practice apparatus. We don't waste time on useless theory. We give you the **practical training** you'll need for your start in Radio **—in 12 short weeks.** If you desire code, this requires additional time for which there is no extra charge.

Mail Coupon Today for All the Facts

H. C. LEWIS, Pres. **RADIO DIVISION** **Founded 1899**

Coyne Electrical School

500 S. Paulina St., Dept. 38-8H, Chicago, Ill.

ELECTRIC REFRIGERATION AIR CONDITIONING AUTOMOBILE ELECTRICAL WORK

Instruction now included at no extra cost. Here is your opportunity to learn these valuable allied lines without extra tuition charge.

PART TIME EMPLOYMENT TO HELP YOU "EARN WHILE LEARNING"

If you are short of money and need part time employment to help pay for your room and board while training, my Employment Department will help you get a part time job.

GET THE FACTS

Don't let lack of money prevent your sending in the Coupon. Mail the Coupon today and I Will send you the Big FREE Coyne Book full of FACTS.

H. C. LEWIS, Pres.
Radio Division, Coyne Electrical School
500 S. Paulina St., Dept. 38-8H, Chicago, Ill.

Dear Mr. Lewis:—Send me your Big Free Radio Book, and all details of your "Pay After Graduation" Plan **including valuable instruction in Electric Refrigeration, Air Conditioning and Automobile Electrical Work.**

Name..

Address..

City................................*State*..............

Dealer's Net Price, F. O. B. Greenwood, Miss. Size 7½ x 12½ x 16½ **$139.50**

Also available in smaller case for the radio man who does not care to carry spare parts, tubes, etc., in same unit.

Again Acclaimed "SUPREME"

Another RMA Show—Another Triumph

Standing the acid test of daily actual use in the hands of thousands of radio engineers and service men the country over, meeting every service requirement in a constantly changing field, the

SUPREME
Radio Diagnometer
MODEL - 400 - B

remains

UNAPPROACHED! UNCHALLENGED!

UNCHANGED! —except for the addition of Pentode Testing to its almost limitless servicing functions!

★ Again the SUPREME DIAGNOMETER MODEL 400-B at the RMA Show was readily conceded by engineers from all parts of the globe to be the unrivalled leader in the field of radio service instruments. Radio dealers, manufacturers, technicians and service men, in ever-increasing numbers are demanding the SUPREME DIAGNOMETER because:

1. Its unique design is outstanding a guarantee against obsolescence.
2. Its rugged construction prevents break-down insuring accurate operation at all times.
3. It is the one service instrument capable of making every necessary test—none other so complete.
4. It pays for itself in a few months—faster and more accurate servicing means greater profits.
5. It provides maximum efficiency with utmost simplicity —pictorial diagrams of every test supplied with each instrument.
6. It creates satisfied customers—provides a laboratory method for the solution of every service problem.
7. It provides the greatest amount of test equipment in the least possible space at the lowest cost per service unit—The service units contained in the Supreme Diagnometer, if purchased separately, would cost many times its price.

And Now!
A Revolutionary Set Analyzer

25 Testing Instruments in 1

All Readings on One Meter—Only One Meter to Read

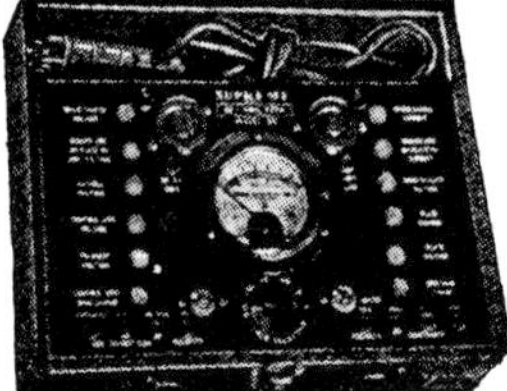

SUPREME SET ANALYZER MODEL 90

Dealer's Net Price, F. O. B. Greenwood, Miss. Size 4¼x9¼x11¼. Shipping weight 6 pounds.

$78.50

Responding to demand for a simple, yet more flexible and complete set analyzer, the "Supreme" Model 90 SET ANALYZER is offered to the Radio service world. Simplified to only one meter of unique design never before available in any but the most expensive laboratory equipment, it measures both A. C. and D. C. voltages up to 900 volts, and currents up to 300 milliamperes, and provides many readings and tests never before attempted with any other similar instrument. You owe it to yourself to compare "Supreme" advantages with any other set analyzer you may be considering!

Most good distributors carry the complete line of Supreme products in stock, including the Model 50 Tube Tester, Model 10 Ohmmeter, Model 17 Tube Checker, as well as the Diagnometer and Set Analyzer. If yours cannot supply you, use order coupon to the right.

Make your Service Department outstanding Modernize with the SUPREME DIAGNOMETER and watch your profits grow.

SUPREME INSTRUMENTS CORPORATION
GREENWOOD, MISS.
Distributors in All Principal Cities—
Service Depots in New York, Philadelphia, Pittsburgh, Chicago, Kansas City, Seattle, Toronto, San Francisco.

Supreme Instruments Corp.
372 Supreme Bldg.,
Greenwood, Mississippi
Please ship Supreme Test Instrument as checked below:

......Model 400-B Diagnometer Net Cash $139.50

......Time Payment Plan—$33.50 Cash and 8 monthly payments of $15.00 each

......Model 90 Set Analyzer Net Cash only $78.50

All prices are F. O. B. Greenwood, Miss. No Dealer's discount.

Date Shipment desired..................

Signed..................

Firm Name..................

....

Street Address..................

....

City..................

State..................

Please give three or more bank or trade references and names of distributors from whom most purchases are made.

SUPREME
Testing Instruments
"SUPREME BY COMPARISON"

THIS ADVERTISEMENT APPEARED IN 1930

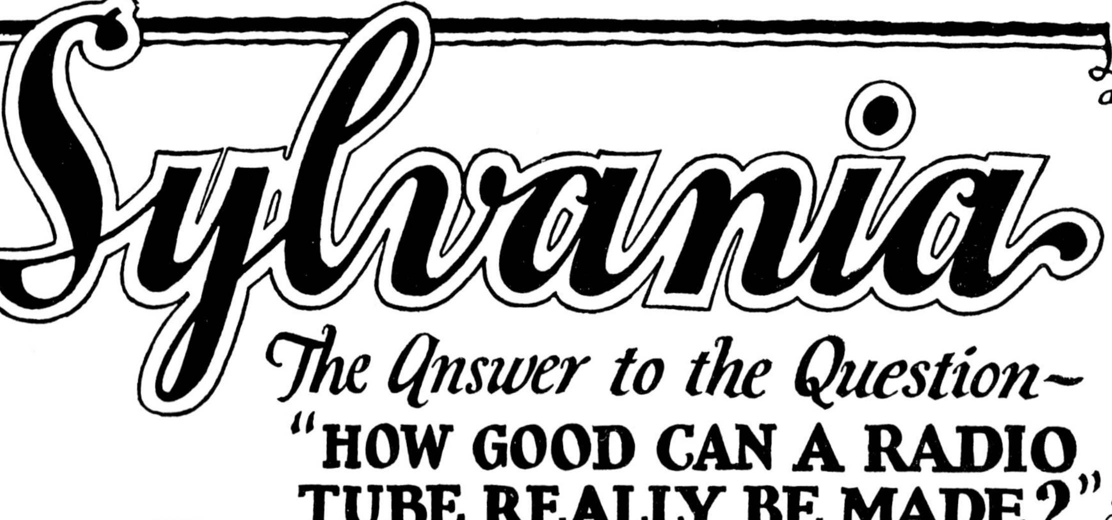

The Answer to the Question—

"HOW GOOD CAN A RADIO TUBE REALLY BE MADE?"

SYLVANIA tubes are not made simply to sell but, first of all to perform. If you have had experience with tubes of questionable quality, this simple statement of a fact will go far in accounting to you for the big difference in quality of tone, sensitivity and volume that you will notice immediately you install Sylvanias in your receiving set.

The fundamental reason for the extra measure of performance that Sylvania tubes deliver is easily understood when you consider the attitude of the makers to their product. They know beyond any question of doubt that the future of radio depends on how good its equipment is made—not how cheap.

Sylvania tubes are made by a close knit organization that has been identified with the manufacture of superior quality electrical products for many years. You are invited to investigate its responsibility through any of the commercial agencies.

To demonstrate to yourself how good a radio tube really can be made, compare Sylvania performance with that of any other tube you may hold in high esteem. Both the manufacturers and the dealers guarantee of complete satisfaction is an obligation that will be upheld under any and all conditions.

We Invite Dealers

—who recognize the relationship of quality to profits to ask us about the Sylvania franchise and the better business building plan. It will be worth your while to investigate regardless of your present connections.

Types: S 01-A, S 99 Large Base, S 99 Small Base

SYLVANIA PRODUCTS CO.
Emporium, Pa.

THIS ADVERTISEMENT APPEARED IN 1925

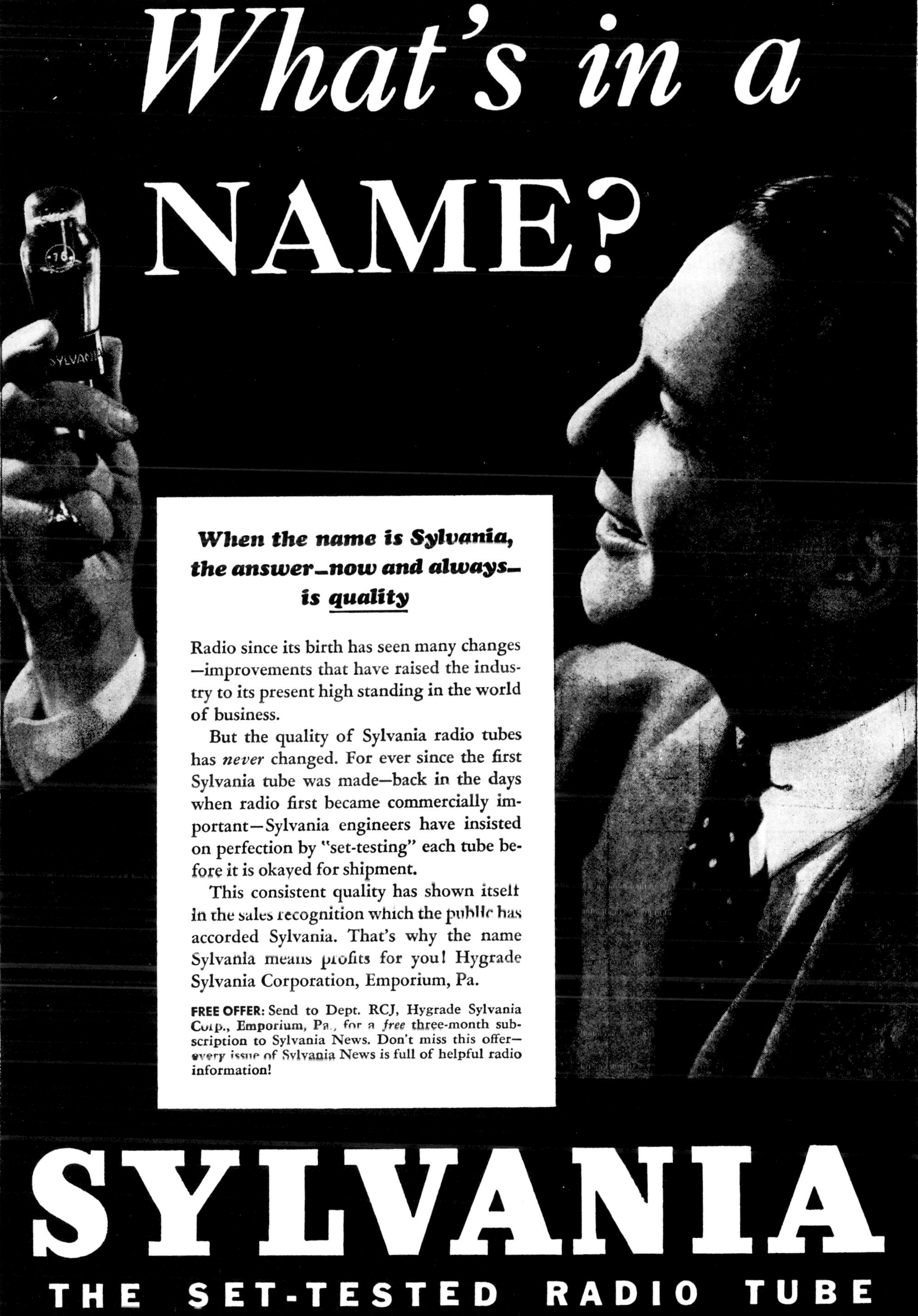
What's in a NAME?
SYLVANIA
When the name is Sylvania, the answer—now and always—is quality
Radio since its birth has seen many changes—improvements that have raised the industry to its present high standing in the world of business.
But the quality of Sylvania radio tubes has never changed. For ever since the first Sylvania tube was made—back in the days when radio first became commercially important—Sylvania engineers have insisted on perfection by "set-testing" each tube before it is okayed for shipment.
This consistent quality has shown itself in the sales recognition which the public has accorded Sylvania. That's why the name Sylvania means profits for you! Hygrade Sylvania Corporation, Emporium, Pa.
FREE OFFER: Send to Dept. RCJ, Hygrade Sylvania Corp., Emporium, Pa., for a free three-month subscription to Sylvania News. Don't miss this offer—every issue of Sylvania News is full of helpful radio information!
SYLVANIA
THE SET-TESTED RADIO TUBE

The New CROSLEY Variable Condenser

"Better — Costs Less"

MODEL "C"

The principle of this instrument needs no introduction Thousands of the Crosley Model "A" Condensers have been sold with uniform satisfaction The Model "A" is conservatively rated at .0005 Mf. The new Model "C" is conservatively rated at .0012 Mf.

While the Model "A" was made with wood frame and laminated wood plates the new Model "C" has ground porcelain plates with die cast frame. It is as efficient a piece of apparatus as you could desire. For tuning C.W. and for power transmission it cannot be equalled. It is tested on one thousand volts before shipment. No body or hand capacity. Low resistance due to absence of spring contacts, copper plates, brass binding posts, etc. We call it the "sensation" of radio—"Better—Costs Less".

Every CROSLEY VARIABLE CONDENSER is GUARANTEED to give absolute satisfaction or money refunded.

The CROSLEY VARIABLE CONDENSER is now made in three styles:

Model "A" with wood frame and laminated wood plates.

Model "B" with wood plates and die cast frame.

(Both models have the same capacity—.0005.)

Model "C" as illustrated—capacity .001—porcelain plates, die cast frame, etc.

Prices as follows:

	Without knob. and dial	With knob and dial	With knob and dial mounted in cabinet
Model "A"	$1.25	$1.75	$2.50
Model "B"	1.75	2.25	3.00
Model "C"	2.25	2.75	3.50

CROSLEY V-T Socket -- 60c.

"Better -- Costs Less"

The biggest selling socket on the market. Practically unbreakable. For either base or panel mounting. Made of one piece porcelain. no metal shell, hence no ground hum. Better—and costs only 60c

Harko Radio Receiver

The most compact and complete efficient crystal receiving outfit on the market Will tune from 200 to 600 meters bringing in spark, voice and music with amateur antenna. A wonderful little instrument. Price complete with battery, etc., **$9.00.** One thousand ohm single head set, 125 feet antenna wire regulators, etc., **$6.00** extra. Complete outfit **$15.00**

Crosley Cabinets

We build a full line of cabinets in gum, mahogany or quartered oak and furnish them with genuine formica panels. Prices on cabinets range from ....**$2.50 to $10.50**

Crosley Detector Units

Furnished completely wired and mounted as shown in illustration or in knocked down form. Price completely assembled **$7.50.** Price of all parts including formica or other panel of high grade dielectric composition, not drilled **$6.00**

Harko Senior Radio Receiver

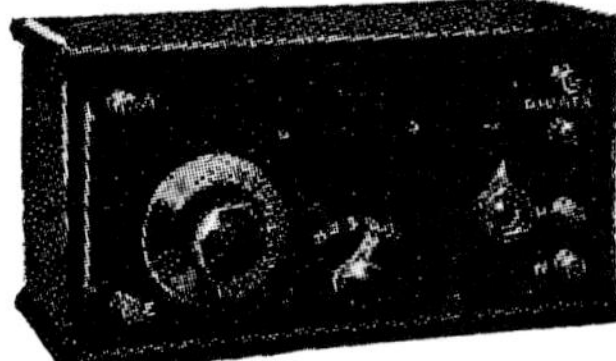

Complete tuner and audion detector assembled on a formica or other high grade dielectric panel, mounted complete in mahogany finish cabinet. Range, 150 to 600 meters, non-regenative hook-up. Price without battery, tube or phones **$16.00**

Crosley Variometer Parts

Consists of two stators, one rotor, the necessary hardware as shown in illustration. Shaft for knob and dial is 3/16" diameter. The wood parts are furnished either in poplar or mahogany. Price of Variometer parts, using poplar wood, **$1.50.** If wood parts are made of mahogany **$1.75**

Crosley Rheostats

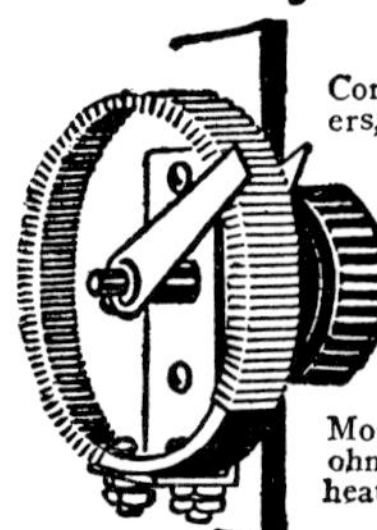

Complete with knob, pointers, etc., as shown in illustration.

Model "A"—overall diameter 1⅝". Resistance 7 ohms, one ampere without heating. Price 60c

Model "B"—Resistance 4 ohms, 3 amperes without heating. Price $1.25

Crosley Two Step Amplifier

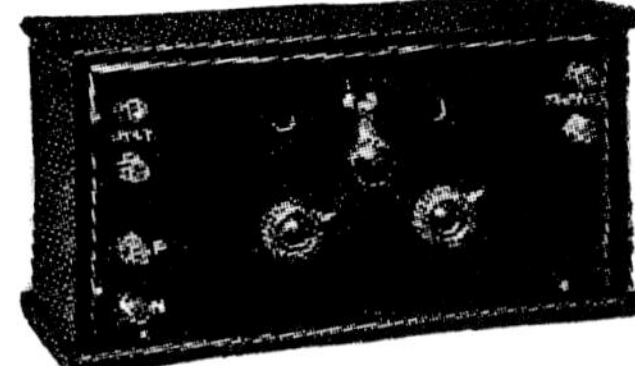

Complete with amplifying transformers, sockets, rheostats, switch, binding posts, etc., mounted on formica panel in mahogany finished cabinet. Price complete as shown in illustration **$25.00**

Crosley Vario-Couplers

Consists of formica tube, rotor and brass hardware. Price, complete as shown in illustration not wound or assembled, **$1.50.** Stator only 40c

Every article guaranteed to give absolute satisfaction or money refunded. If your dealer can't supply you, send us his name and order direct.

Dealers and Distributors. Every item shown above should be in your stock. Write for proposition.

CROSLEY MFG. CO. Radio Dept. R-7 CINCINNATI, O.

THIS ADVERTISEMENT APPEARED IN 1922

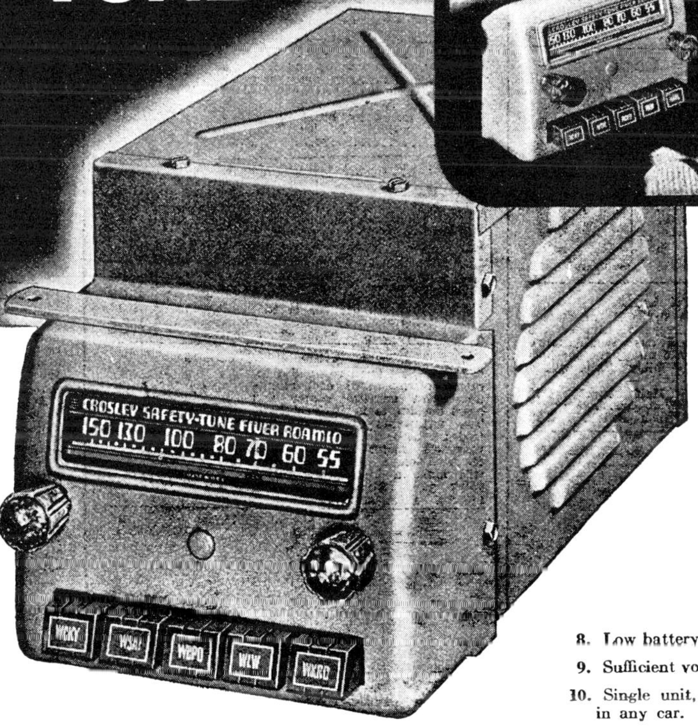

Buy Reliable Radio Apparatus

ORDER DIRECT FROM THIS PAGE

Wireless Transformers

Thordarson Type R

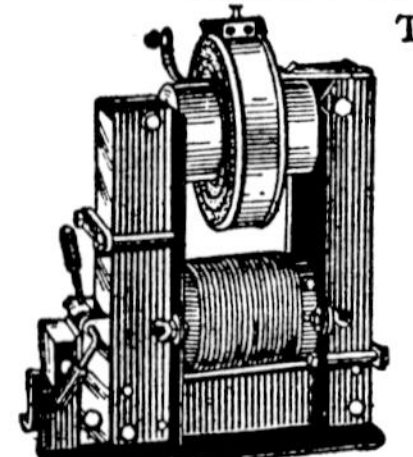

For use on 108 to 115-volt alternating current, 60 cycles. Generally accepted as the standard transformer for amateur transmitting. Provided with adjustable magnetic leakage gap which controls primary input giving a wide range of amperage and permitting easy adjustment to point of resonance. No impedance or choke coil necessary in primary circuit. Ship. wt., 35 and 55 lbs.

Art. No.	K. V. A.	Amp.	Sec. Volt	Price
63R630	½	1 to 6	10,000	$21.00
63R632	1	2½ to 14	25,000	38.00

Thordarson Type R. S. Transformer

This type differs from the well known model shown above only in that it does not have the adjustable magnetic shunt. All other features of sturdy, compact construction and correct electrical characteristics are the same. For use on 105 to 120 volt 60 cycle alternating current.

Art. No.	K. V. A.	Sec. Volts	Ship. wt.	Price
63R633	¼	8,000	15 lbs.	$10 00
63R634	½	10,000	25 lbs.	19 00
63R635	1	15,000	35 lbs.	28.50

Kick Back Preventer

63R5358—Prevents high frequency surges from discharging back into power line. A necessity when power transformer is supplied from city mains. Two 1000-ohm resistance rods. Mahogany finished base. Connections of strip copper. Ship. wt., 4 lbs. Price **$4.80**

Variable Transmitting Condenser—Oil Immersed

63R620—An oil immersed variable condenser for use with all makes of transformers up to 1 K. W., 25,000 volts. Phenol fibre dielectric, corrugated aluminum separators allow circulation of oil to keep down heating. Flat aluminum sheet electrodes with rounded corners. Variable in ten steps of .0009 M. F. each from .0018 M. F. to .009 M. F. Especially designed to prevent corona losses and brush discharge. Oil included. Ship. wt., 35 lbs. Price **$25.00**

Wireless Spark Coils

For use on dry cells or storage batteries. Properly adjusted the half-inch coil has a sending range of from 2 to 5 miles, the one-inch coil 5 to 10 miles. Ship. wt., 6 and 8 lbs.
63R5126 — Half inch coil **$3.65**
63R5127—1 inch coil 5.95

Spark Coil Transmitting Condenser

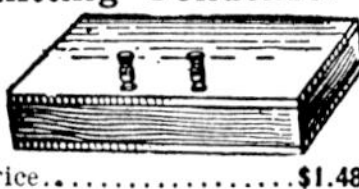

63R5348—Designed for use with spark coil sets, dielectric of 5—size, 5x7 photo plates. Mahogany finished case. Permits working on 200 meter wave. Ship. wt., 3 lbs. Price $1.48

Murdock Improved Antenna Switch

63R5113 — A large, sturdy, well built "change over" switch suitable for use on sets up to 1 K. W. Mahogany finish base. Improved support, copper blades. Fitted with third blade to disconnect receiver when sending. Ship. wt., 2½ pounds. Price **$4.30**

Zinc Spark Gap

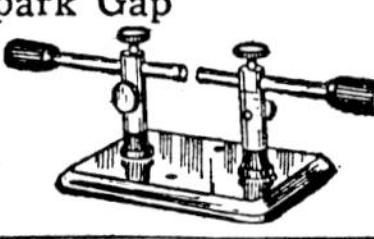

63R5350—For use with spark coil transmitters. Base is moulded composition. Metal parts are plated and polished. Can be used with coils up to 4 inches. Ship. wt., 2 pounds. Price $1.10

Radiator Spark Gap

63R5351—Micrometer adjustment. Electrodes of zinc. Cooling vanes aluminum. Metal parts nickel plated. Base glazed porcelain. Will handle over 1 K. W. Weight, 2 lbs. $2.30

Combined Detector and Two Stage Amplifier

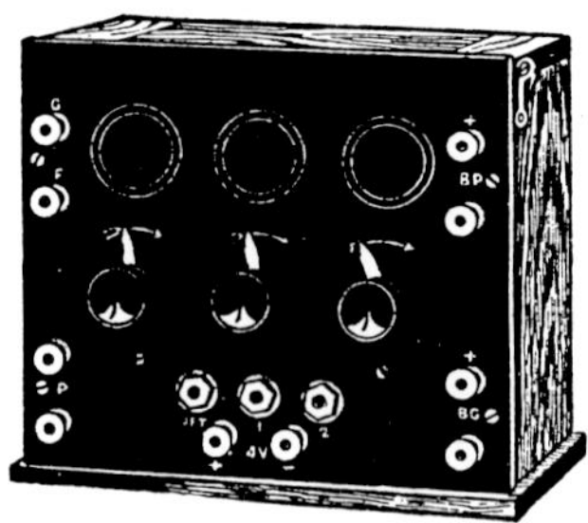

This instrument with the proper accessory instruments has the broadest possible range. So sensitive is it that even with an indoor aerial such as our loop antenna 63R650 messages may be received from stations hundreds of miles distant. Combined with our Regenerative set and two variable condensers and the proper couplers and loaders with a good outdoor aerial, its range is increased to equal the best commercial stations. Has one detector and two amplifier circuits. Standard tube sockets, grid condenser in detector circuit, two amplification transformers, three jacks and one plug so that any desired circuit may be used. Bakelite panel 7½ in. high, 8¾ in. wide. High grade weathered oak finish cabinet with hinged top making interior easily accessible. Binding post connections for batteries and couplers. Binding posts black oxidized finish. No tubes, batteries or phones included. Shipping weight, 18 lbs.
63R615—Price $33.50

Commercial Type Oscillation Transformer

$16.50

63R648—Designed to give wave ranges both above and below 200 meters. Solid copper windings on "Formica" supports 10½ in. diam. primary of six turns No. 3 wire 6½ in. diam., secondary of twelve turns No. 5 wire. Mahogany finished woodwork. Two helix clips included. Ship. wt., 26 pounds. Price $16.50

Murdock Oscillation Transformer

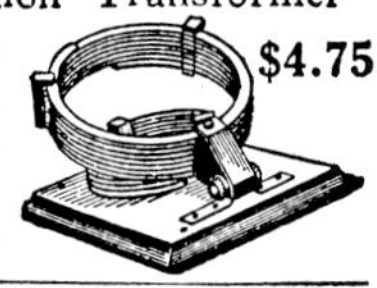

$4.75

63R5155—Permits sharp tuning on 200 meter wave. Can be used on sets up to 1 K. W. Primary and Secondary windings of edgewise wound copper ribbon. Coupling varied by hinge. Shipping weight, 8 lbs. Price $4.75

Universal Spark Gap Motor

63R5624—A rugged high grade motor for spark gaps, running sewing machines, fans, small lathes, buffers, emery wheels, etc. Running idle will make 8,000 R.P.M. Will make about 4,500 R.P.M. with electrode shown below. Will operate on 100-125-volt A. C. or D. C. current. Black enamel finish. Height, over all, 5½ in. ¼ in. shaft, extends ⅞ in. Will develop about 1/20 H. P. Supplied with 1-in. grooved pulley. Ship. wt., 8 lbs. Price **$9.50**

Spark Gap Electrodes

63R5625—Saw tooth rotor, 5¾ in. diam.; of machined cast aluminum with bakelite center and brass bushing to fit ¼-inch shaft. Two adjustable stationary electrodes. This set together with universal motor listed above, mounted on a substantial base will make a high grade rotary spark gap. Ship. wt. 1½ lbs. Price, set **$4.50**

Improved Model Rotary Spark Gap

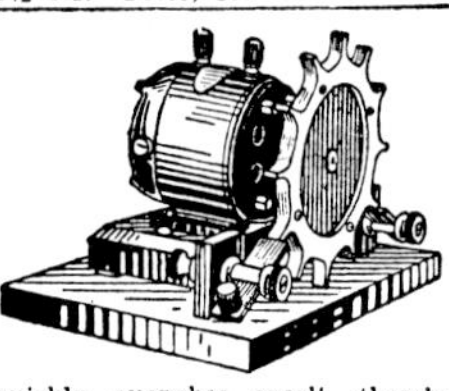

Flat pure copper electrodes avoid pitting. Width of break adjustable. Strong breeze generated by rotary electrode quickly quenches spark, thereby allowing transmission of wave of low decrement. All conducting metal mounted on Formica. Easily handles 40,000 volts without endangering motor windings. Constant steady speed. Ship. wt., 10 lbs. Shipped from stock at Chicago.
263R5142—½ K. W. size. 1/20 H.P. Universal motor. For 108 to 115-volt current. Speed 4,000 R.P.M. Price $14.80
263R5143—1 K. W. size. 1/12 H.P. Universal motor For 108 to 115-volt current. Speed 5,000 R.P.M. Price $18.95

Ground Switch

63R5359—600 volt, 100 ampere double throw single-pole switch on insulating base. For grounding aerial when not in use. Ground wire should be 4 gauge. Ship. wt., 5 lbs. Price **$3.78**

Ground Rod

63R1061—Iron Ground Rod; length, 6 ft. Heavily galvanized. A ground rod is necessary with every Radio outfit to insure a perfect ground contact. Shipping weight, 4 lbs. Price each **43c**

Double Action Wireless Key

63R5356—Double action which makes for speed. Will improve your sending and lends individuality. Large standards, formica knob, heavy silver contacts suitable for use up to 2 K. W. Mounted on durable and heavy formica base. Metal parts nickel plated. Ship. wt., 1¾ lbs. Price **$4.65**

Standard Wireless Key

63R5352—A high grade key made of heavy lacquered brass. Large hardened contact points. Ship wt., 1 lb. Price **$3.48**

Steel Lever Keys

Shipping wt., 14 oz.
63R1739—Leg Key with legs to go through table or desk. Price, each **$1.85**
63R1741—Legless Key to screw to top of table or desk. Price each $1.97

Wireless Practice Set

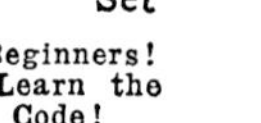

Beginners! Learn the Code!

Anyone learning wireless must know the code. Without it, it is impossible to understand the signals. Send for a wireless practice set and see how easy it is to learn the code. Set consists of a key and buzzer mounted on a polished wood base. Buzzer reproduces accurately the high pitched sounds of wireless stations. Connect a dry battery to the binding posts on the set by means of a short piece of wire, press the handle of the key and buzzing sound will be produced. In a very short time your ear will become accustomed to the various combinations of dots and dashes representing different letters and numerals. Practice until you can understand the signals at the speed sent by average stations and you have completed the most difficult part of wireless telegraphy. A very good way to learn the code quickly is to place two of these sets in separate rooms with an operator at each set and practice sending signals back and forth. Chart included with each set. Base size, 7x4¾ inches. Shipping weight, 3 pounds.
63R1750—Wireless practice set. **$2.10**

Standard Type Transformer

Operates only when connected onto alternating current line of 100 to 125 volts, 60 to 133 cycles. 100 watt Transformer produces 1½ to 24 volts, 150 watt 1½ to 30 volts in steps of 1½ volts each, and in addition can produce constant voltages, so that a number of different voltages can be drawn at the same time. Fitted with 7 ft. cord and attaching plug.
63R1697—Capacity, 100 watts. Ship. wt., 10 pounds. Price **$4.35**
63R1699—Capacity, 150 watts. Ship. wt., 13 pounds. Price **$5.95**

American Ignitor Dry Cell

We guarantee that these batteries will reach you fresh and will give you all the service you have a right to expect of the best 2½ by 6-inch dry cell made. Designed especially for heavy duty work. For ignition work on gas engines, automobiles, motor boats, etc. Average weight, 2 pounds. Size, 2½ by 6 inches. Average initial amperage, 25 to 30 amperes.
63R2501—Price, each **37c**
63R2502—Price, per dozen $4 30

CHICAGO, ILL. ST. PAUL, MINN. KANSAS CITY, MO.

Send Your Order to House Nearest You

THIS ADVERTISEMENT APPEARED IN 1921

HUGO GERNSBACK, Editor-in-Chief

N. H. LESSEM, Associate Editor
C. P. MASON, Associate Editor

R. D. WASHBURNE, Managing Editor

J. T. BERNSLEY, Technical Editor
ROBERT EICHBERG, Research Editor

THOS. D. PENTZ
Art Director

CONTENTS-March, 1938, ISSUE

Volume IX Number 9

Editorial: 50 Years of Radio .... *Hugo Gernsback* 541

The Growth of Broadcasting .... *N. H. Lessem* 542

Progress of Radio Receiving Equipment .... *Robert Eichberg* 546

Origin and Growth of Radio Servicing 551

A Chronological History of Radio .. *J. T. Bernsley* 552

Reminiscences of Old-Timers 556

The Story of Amateur Radio 562

Radio Parts of Yesteryear 564

Development of Sound Equipment 568

The Old E. I. Co. Days .... *Hugo Gernsback* 572

Mileposts in Television 576

Fleming's Valve — and Up 580

Famous Early Radio Circuits 584

Some Old-Time Radio Stations 588

Radio in the Future 590

Predictions and Fulfillments 592

New Circuits in Modern Radio Receivers — No. 6 .... *F. L. Sprayberry* 595

The Radio Month in Review 596

How to Make the *Radio-Craft* Super-Deluxe 30-Tube Set — Part VI 598

New Tubes Introduced to Radio Men in 1937-'38 .... *R. D. Washburne* 600

Book Reviews 612, 617

Super-Regeneration—in 1922 614

When the Neutrodyne Made Its Bow 615

Early Tube Experiments 616

A $750, Kc.-Calibrated Tuner—in 1927! 628

Credits 655

Published by Radcraft Publications, Inc. Publication office: 29 Worthington Street, Springfield, Mass. Editorial and Advertising Offices: 99 Hudson Street, New York City. Chicago Advertising Office: RADIO-CRAFT, 520 North Michigan Avenue, Chicago, Ill.

RADIO-CRAFT is published monthly, on the first of the month preceding that of date; subscription price is $2.50 per year in U. S. and Canada. (In foreign countries, $3.00 a year to cover additional postage.) Entered at the post office at Springfield as second-class matter under the act of March 3, 1879.

Foreign Agents

London—Gorringe's American News Agency, 9A Green St., Leicester Square, W. C. 2, England.
Paris—Messageries Dawson, 4 Rue Faubourg, Poissonniere, France.
Melbourne—McGill's Agency, 179 Elizabeth St., Australia.
Dunedin—James Johnston, Ltd., New Zealand.

APRIL RADIO-CRAFT

The annual SPECIAL ELECTRONICS NUMBER will contain not only many articles of exceptional interest to experimenters and practitioners in the electronics field but also numerous articles on other allied fields of radio; and of course, as usual, the various monthly departments. This issue will tell you about new developments in electronic music, test equipment, sound apparatus, and radio in general. Also in April "*R.-C.*" will appear an unusually informative article for the Service Man; and another on a 10-tube all-wave receiver for the set builder; a third on an "expressor" for Public Address specialists; and, finally, several departments which due to lack of space do not appear in this March issue. All these and much more will be found in the forthcoming April issue of *Radio-Craft.*

"Takes the Resistance out of Radio"

Editorial Offices: 99 Hudson St., New York, N. Y. HUGO GERNSBACK, Editor Vol. IX, No. 9, March, 1938

50 Years of Radio

An Editorial by HUGO GERNSBACK

WHEN Heinrich Hertz performed his epoch-making experiments during the years 1887-'88 in Frankfurt, Germany, he little dreamt what a tremendous force he was about to loose upon humanity.

It is unfortunate that this illustrious scientist has had so little recognition, and, indeed, the proverbial man in the street does not even know his name. Most people think that Marconi is the man responsible for radio. Of course, Marconi *is* responsible for the practical application of radio, but it was Hertz who laid the ground work; and it is a fact that he did all the research work upon which modern radio is based. As time goes on, we must marvel more and more at the tremendous insight which Hertz had into radio physics, for many of the principles which he discovered are being used only of late.

Thus, for instance, short waves were first used by Hertz only to be discarded when radio got under way, and but lately "rediscovered." The art of radio advanced along orthodox lines the same as any other art, and the process was just as painful, but as regular, as in other arts. It is always an easy matter to look back upon what has gone before, and laugh at the mistakes which were made when we were still groping in the dark. From our elevated height in looking backwards over 50 years of radio, it would be easy to point out the mistakes, the fallacies and the blunders which were committed during the entire development stage of the first half-century of radio.

The pity of it is that the hard work which went into the art of radio, and which contributed so largely to its present state of development, is not recognized by most of us in this latter day. Many are the heroes of radio who are buried in darkness, and who have had little recognition, if any. Indeed, it perhaps would be pointless to give a list of the men whose illustrious work made radio what it is today, in an article of this type. Ours is an impatient world which is not concerned with names and personalities, but which wishes only to see the results of their labors. This is most unfortunate, but it has ever been thus with pioneers, and those who have slaved and given their best to make a new art possible.

As a small monument to those who gave their effort and best years to the development of radio, this issue of *Radio-Craft* has been dedicated.

Radio would not be what it is today without those valiant souls who stayed up night after night, and often wrecked their health, but who gave us and posterity present-day radio.

It is to these unsung, uncrowned men, who deserve some recognition (even if it is only in reading their name for the first time, as perhaps you will when perusing this JUBILEE SOUVENIR NUMBER of *Radio-Craft*) that this issue is dedicated.

While it has not been possible to give as much space as we had hoped to all radio pioneers, we believe we have included in this issue of *Radio-Craft* almost every radio personality whose work has gone into modern radio.

And I, for one, believe and sincerely hope that if you are at all interested in radio, you will find time to peruse the entire issue, if for no other purpose than to engrave on your memory those names of radio men, without whom radio would not have developed into the miraculous art which has become so tremendous a part of our daily lives.

And it is to those men, to their tireless efforts, and to their work, to whom I dedicate this, the JUBILEE SOUVENIR NUMBER of *Radio-Craft*.

THE GROWTH OF

HOW *modern broadcasting grew from the experiments of a Pennsylvania amateur, conducted shortly after the World War, to a mammoth industry is told in the following article.*

N. H. LESSEM

Dr. de Forest and wireless phone, 1907

Aerophone Tower of Radio Telephone Company of New York atop the Terminal Building near 42nd St.—about 1908

Dr Frank Conrad's early radiophone experiments began in a garage at the rear of his home, 1919

This Pittsburgh Sun advertisement was instrumental in starting radio broadcasting over KDKA Sept. 29, 1920

Air Concert "Picked Up" By Radio Here

Victrola music, played into the air over a wireless telephone, was "picked up" by listeners on the wireless receiving station which was recently installed here for patrons interested in wireless experiments. The concert was heard Thursday night about 10 o'clock, and continued 20 minutes. Two orchestra numbers, a soprano solo—which rang particularly high and clear through the air—and a juvenile "talking piece" constituted the program.

The music was from a Victrola pulled up close to the transmitter of a wireless telephone in the home of Frank Conrad, Penn and Peebles avenues, Wilkinsburg. Mr. Conrad is a wireless enthusiast and "puts on" the wireless concerts periodically for the entertainment of the many people in this district who have wireless sets.

Amateur Wireless Sets, made by the maker of the Set which is in operation in our store, are on sale here $10.00 up.

—West Basement

Epochal Nov. 2nd Broadcasting over KDKA. Station view of personnel during broadcasting of Harding-Cox election returns, 1920

Dr. Conrad's "garage" experimental broadcasting station 8XK prior to 1920

A Broadcast over WJZ, Newark, about 1921-22

N ORDER TO TRACE broadcasting's growth, one must first arrive at a definition of the term. Do we include the experiments of Hertz and Loomis? Did broadcasting commence with Marconi, de Forest or with Conrad's transmissions?

For the purposes of this article, a rather definite distinction must be drawn between bona-fide broadcasting and the mere transmission of signals (be they voice, music or Morse) by wireless means.

A DEFINITION

In its exact sense, the term "broadcasting" includes only such signals as are sent out by means of radio waves *for general reception.* A signal directed from one point to another in code so that it is unintelligible except to a certain or restricted few, is certainly not "broadcast" as the average individual or listener-in would define it.

His conception of "broadcasting" would refer more specifically to speech or music sent out for the information, entertainment or education of the general public,—consequently, broadcasting as America knows it today may be said to have had its true birth early in 1907—when de Forest constructed the first means of modulating an arc transmitter with voice impulses and began *sound* broadcasting from atop the 12-story Terminal Building at 42nd St. and Park Ave., in New York City.

So successful were the results that de Forest's company soon received an order for an installation on a private de luxe yacht, which was quickly followed by the sale of 24 complete radiophone stations to the U. S. Navy for installation on battleships. It is interesting to note that the latter units were sold with a positive guarantee of 5 miles reliable range, although actual operating tests showed them to be consistently good over a distance of 26 miles.

In 1908, de Forest went to France and installed his radiophone transmitter atop the Eiffel Tower, from where he broadcast phonograph music. Reports of this test showed that he was heard as far as 500 miles,—a feat which a few more modern stations could not duplicate today.

In January 1910, the De Forest Radio Telephone Company began a series of broadcasts, employing actual singers and grand opera stars to provide the programs. The first grand opera broadcast was a "double-feature", *Pagliacci* and *Cavalleria Rusticana,* and Caruso

BROADCASTING

The major steps in the transformation of "wireless" to "radio", ethereal chaos to orderly regulation—with side excursions into studio technique are here chronicled.

sang in the latter. These efforts and the talent were wasted, however, for it was too soon for broadcasting. Only amateur wireless operators had receiving sets and they were very few in number at that time.

And so for a while broadcasting lay dormant, until the day when it would catch the public's imagination like wildfire and create an insatiable demand for receivers and parts. In the interim, during the early part of the World War to be exact, the A. T. & T. Company was conducting a series of tests and making developments in wireless telephone transmissions, for the purpose of providing facilities for long-distance telephone communication without wires. They had purchased the sole rights to de Forest's audion tube, perfected modulation systems (the vacuum tube as an oscillator was already discovered by both de Forest and Armstrong), and were finally successful in broadcasting telephone speech clearly and reliably over long distances. By the latter part of 1915, telephone conversations were broadcast across the Atlantic,—to illustrate how far their progress had gone. These early researches were destined to be the basis for future broadcast station design,—without them, broadcasting would have certainly been later in coming to the fore.

AN INTRODUCTION—KDKA

In 1919, Dr. Frank Conrad—amateur and Westinghouse engineer—began to broadcast programs from his station which was located in the garage at the rear of his house in Pittsburgh. The programs consisted mainly of phonograph records, but by this time so many amateurs were in existence and so many were their friends who would come over to listen to the music, that Conrad began to receive avalanches of mail. Even newspapers began to publicize Conrad's stunt, listing at times the hours of broadcast. So great became the enthusiasm of nearby radio amateurs that the advertising department of a leading Pittsburgh department store got a bright idea for selling some Army surplus apparatus which it had on hand. "Buy one of these World War sets," the ads proclaimed, "and you will be able to hear Dr. Frank Conrad."

Ideas breed more ideas, and when Westinghouse officials saw the advertisement, a startling idea struck them:—"Why not build a transmitter right at the factory?", they thought, and proceeded to do so. *By the middle of 1920,*

W.B.Z.'s Combination Station and Studio, Sept. 10th 1921

W.D.Y. Between Aldene and Roselle Park, N.J. 1921

RCA's WDY Antenna System, 1922

Radio-Craft's Editor-in-chief Hugo Gernsback lecturing over WJZ. Listening audiences, an estimated 100,000. Note "Mike"—an ordinary transmitter—1922

WGY Broadcasting Studio Schenectady, N.Y. 1922

William Jennings Bryan's Point Breeze, Presbyterian Church Broadcast at KDKA, March 12, 1922

Early W.E. Equipment at WHK-WCLE, 1922

KDKA's first formal studio, inside one of the Westinghouse factory buildings, was soundproofed with burlap—1922. Note counterbalanced mike stand. W.E.

Guglielmo Marconi witnessing a demonstration by E.W. Rice, Jr. (deceased) of an unusual G.E. microphone—1922. G.E.

it had been assigned call letters, and the little shack perched atop the 9-story factory was known as KDKA.

Then station officials got in touch with the executives of the *Pittsburgh Post*, and *Sun*, making arrangements for the returns of the Presidential election to be broadcast. On November 2, 1920, the station broadcast the bulletins and finally the news that Warren G. Harding was the country's newly-chosen chief executive.

Not recognizing a competitor which was later to cause them untold worry, various newspapers hailed this achievement. Some 2,000 of them, in fact, were so impressed by the performance that they commenced printing KDKA's programs, as a matter of reader interest. In doing so, they gave broadcasting its first foothold, which it has consistently enlarged.

Once started, broadcasting continued to spread like living flame, but before turning to contributions made to the art by other stations, let us tip our hats a few more times to KDKA. This station broadcast the first radio church service —and think of that when you get an airful next Sunday.

When you tune-in the speeches at a banquet, recall that KDKA was the first station to handle a broadcast of that sort. The same also holds true for the first remote pick-up, the first presidential inauguration, sporting event, baseball scores, time signals, market reports, etc.

THE "A" AND "B" RANGES OF 360 AND 400 METERS, RESPECTIVELY

In broadcasting's early days, stations were licensed indiscriminately, without regard for the kilocycles and the interference that many stations would create when crowded together in a narrow channel. The operating frequency was at first 360 meters or (A) range, but soon broadcasting mushroomed from one station in 1920 to 600 stations in 1922, and the Department of Commerce was forced to include a 400-meter channel, or (B) range for the higher-power (and quality) stations.

At that there was considerable interference (even though

Warren Cox, WHK's founder, at microphone. Note megaphone for directing sound waves to telephone-stand microphone—1922. MCH

WGY's short-wave remote pick-up portable transmitter—1923. G.E.

KDKA's Pioneer pick-up short-wave transmitter—192–

Full-fledged broadcasting over WJZ—1923. W.E.

Typical outdoor religious broadcast over KDKA. Note Tomato-Can Microphone 1923-'24. W.E.

W.E.

BROADCASTING

this number of stations was spread out over the country),—so that at that time we find the broadcasters themselves drifting slightly away from their assigned frequencies to minimize or prevent the interference which they were creating. By 1923, the broadcasting range had changed to 200 to 550 meters, and none too soon, for by 1924 a total of 1,400 broadcasting stations were in existence.

IMPORTANT CONTRIBUTIONS TO BROADCASTING

From the standpoint of reception, a contribution of WLW's is of almost equal importance, and though other developments antedate it, it merits consideration here. It was the first remote control transmitter.

One early broadcasting peculiarity was that the transmitter and studios of any station were housed in the same building—the closer together the better. While convenient, it was not sound practice from the viewpoint of either the engineering or the program departments, for if the station were located in the heart of a city, the large neighboring buildings absorbed much of the radiated energy; if it were located anywhere else, it was hard to persuade top-notch artists to journey to it.

Performers were unpaid in the early '20's, and there was scarcely a more diverting sight than a taxi-load of hilarious and high-salaried opera stars bound for a station, elated over the free ride. The microphone into which they sang upon their arrival looked like an antique phonograph horn.

Returning from this digression, one must bow to the telephone company rather than to the broadcasting stations themselves for a step which may be considered to rank at least third in importance: the development of radio program transmission lines and, subsequently, high-frequency lines.

Standard telephone lines were at first the sole means of carrying programs from or to remote points. As such lines were designed to transmit only the human voice, they carried only a comparatively narrow band of audio frequencies,

(Continued on page 606)

KDKA's marble-panel transmitter-1924

At Syria Mosque, one of KDKA's more than 40 pick-up points in Pittsburgh, in 1926

Section of power room, NBC's Radio City headquarters. Power control desk at right background -1938

Calvin Coolidge's Lincoln's Birthday address over WEAF and WJZ. First national-wide hookup-1924

WHK's Studio Control room-1925

Station 2XAR (WJZ), RCA's high-power experimental broadcaster, Bound Brook, N.J.-1928-30

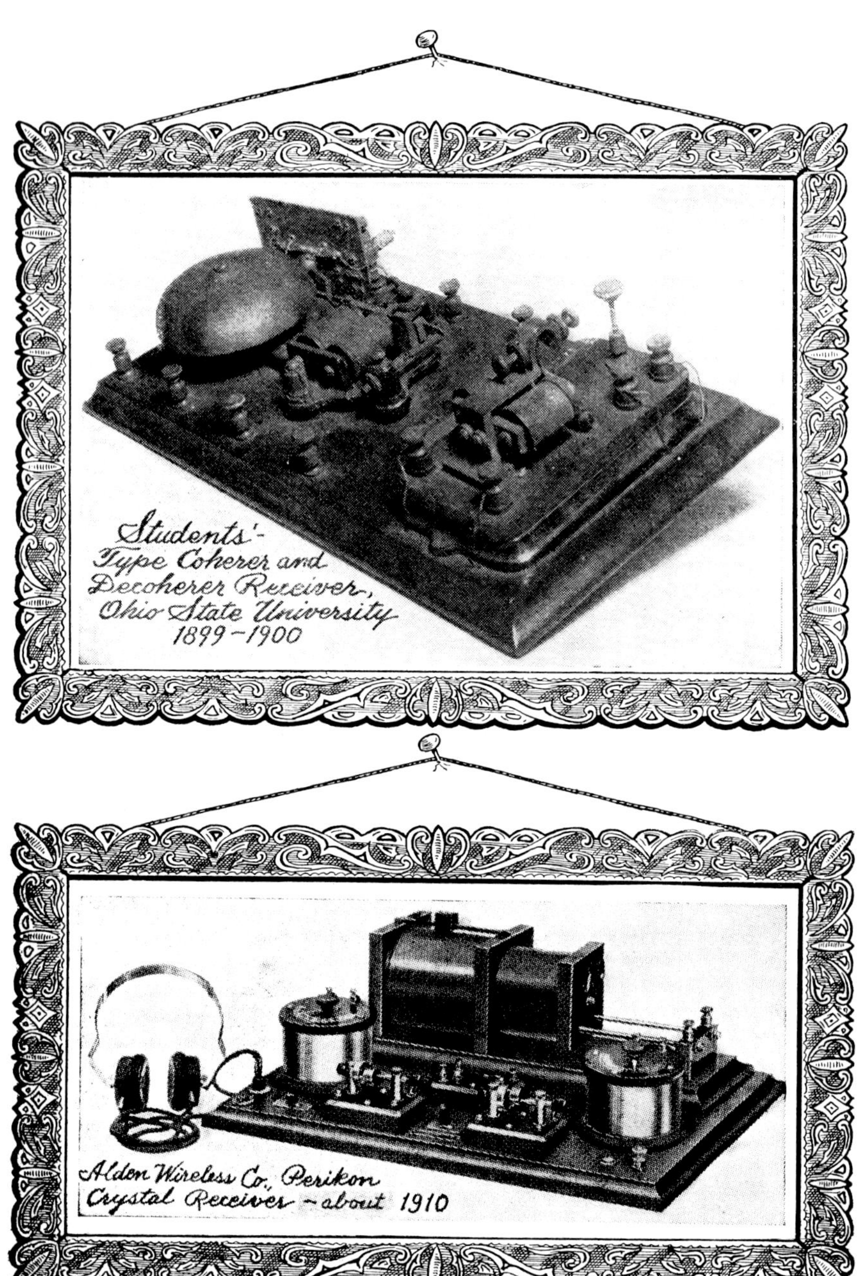

Students'-Type Coherer and Decoherer Receiver, Ohio State University 1899-1900

Alden Wireless Co., Perikon Crystal Receiver—about 1910

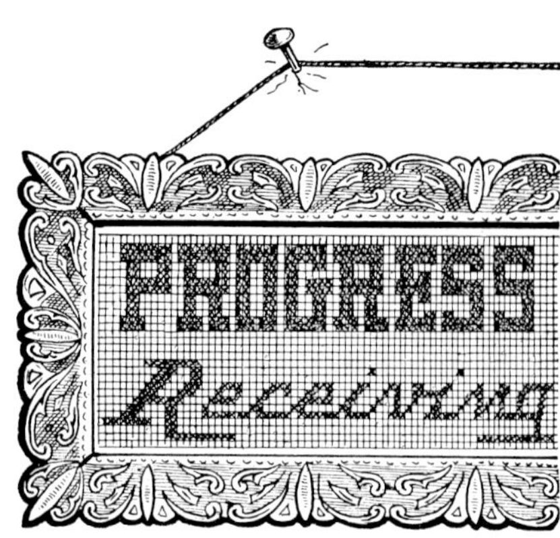

ROBERT

REMEMBER when it was hard to tell speech from music—and static might readily have been mistaken for either, so faithful was reproduction? Ah, my ancient ones, those were the "good old days!"

Few, if any, receivers were sold complete, even in the crystal models. You bought the set, the headphones or "phones" (there were no loudspeakers—they're modern) and the antenna kit separately, and hooked them together as dictated by the instruction sheet and your own conscience, with the latter predominating. The combination cost you more than a 2-band A.C.-D.C. midget does today, for you paid from $12.00 to $25.00 for the set, $2.00 to $5.00 for the antenna outfit and spent whatever you had left for the phones—and walked home. Phones might even cost you more than the set itself, if you wanted "DX".

Having a factory-built receiver was a mark of distinction, for most people of radio's 'teens and early 'twenties were graduating into radio via the oatmeal-box route. And then when sets which really used tubes came out—!

The early single-tube sets were, in effect, a diode rectifier replacing the crystal, which had been a marked advance over the coherer. Though technically "radio," they were not what one generally has in mind when considering early sets. The first sets used by the average listener of today were the rather highly-developed crystal sets, and the sets using such early 3-element tubes as the WD11 and the UV201.

United Wireless Co. Type E Tuner With Carborundum Crystal - 1912

Early tube sets were only slightly more expensive than crystal sets and but little better. Some of the crystal outfits contained a buzzer (which, plus its usual clearly audible buzz, produces—like a local radio station—strong radio signals or buzzes), a battery and a button, to enable the user to find a sensitive spot on the crystal. He pressed the button and listened for the buzzes (unless you had muffled the *audible* buzz, you sometimes couldn't hear the *radio* buzz!) in his phones, while he felt around on the surface of the galena crystal with the metal "cat-whisker" (whisker-like wire) contact until he had located a sensitive spot. If some one then walked across the room, the contact often jarred out of position, making a new search necessary. There was quite a lot of advertising on crystal detectors with cat-whiskers which would stay in position. Iron pyrite detectors were also favored for that characteristic—and fixed crystals, in which a sensitive spot was located and the cat-whisker permanently attached in position, sold for $1.00 to $2.50 each. Carborundum detectors, energized by flashlight batteries, had the same appeal.

No jarring out of position bothered the users of the early tube sets, but they did have to ask the family please to sit still while a distant station was being received. Tubes were "microphonic," and a footstep in a different part of the house often came through the phones with a sound much like Major Bowes' gong, only duller. To overcome microphonics, lead caps were often placed on the tubes,

EICHBERG

The first tube sets were neither sensitive nor selective. They were, for the most part, single-circuit sets—merely a condenser-tuned coil connected between antenna and ground, with its antenna end connected to the grid of the tube through a grid-leak and condenser; its ground connected to the negative side of the filament. The phones were simply hooked between the tube's plate and the "B" battery's positive terminal.

Then came regeneration, bringing joyous chaos in its wake. All that was done was to introduce a "tickler" coil in series with the tube plate. This effectively reduced the operating resistance of the grid circuit, permitting the tube to operate at far higher efficiency. However, it often caused circuit oscillation.

As a result, each set became a miniature transmitter, sending its howls for blocks. Radio listeners did this quite deliberately, for they soon learned the "dead beat" method of tuning, through making the set howl, then tuning the receiver; the pitch of the howl changed as a carrier wave was approached and the stations came in with a sound much like "bloop" and *ergo*, the sets became known as "bloopers."

Even with these regenerative sets, headphones were the customary adjunct, though on powerful local stations, one might lay the phones on the table and hear the program, such as it was. Myriad numbers of crushed and aching ears rose up and demanded a means of listening to more remote stations with the phones on the table—so audio amplifiers and loudspeakers began to creep timidly toward the market place. Soon you could buy, for about the cost of your set, a 2-stage amplifier to add to it. Then you could put your phones in a cut-glass fruit bowl and "let the whole family listen-in," as the ads put it. Incidentally, there were two schools of thought during this period. One claimed that a wooden chopping bowl gave best results; the other, that cut-glass was more brilliant.

Soon the bowl no longer cheered, and people demanded something better. It was offered to them in the form of a cast-aluminum horn with a pair of arms over which the phones could be clasped. The horns sold for $12.50 when they first appeared, but soon cheaper ones, made of papier maché, came out, claiming that they were better because they did not resonate so freely.

About this time there were also phonograph adapters. These did not permit you to play your phonograph through your radio set; conversely they permitted you to play your radio set through your phonograph. All the élite owned phonographs in those days, and these instruments had well-engineered tone-arms and horns. The adapters permitted the sound from the radio set's phone unit to come through the phonograph's horn. If it sounded almost as well as a record, that was fine, for the early sets were not all that might have been desired from the standpoint of fidelity. Bass notes were almost lacking as were, of course, harmonics of treble frequencies. Speech came through harshly; music with tinny tone. The way one judged a set was to tune-in a piano selection and if all the notes were

PROGRESS OF RADIO
Receiving Equipment

C.R.L. Paragon Regenerative—about 1921

Pacent Duolateral-Coil All-Wave Crystal Set—1921

First Radio Phonograph Console Combination With Loop Ant. Designed by—Hugo Gernsback—1921

Westinghouse "Radiola Jr." Crystal Set—1921

G. E.'s First Radio Receiver—A Crystal Set—1921

G. E.'s Bedside Loop-Antenna Receiving Set—1922

Radiola's Famous Long-Range Combination, R.F. P.A. and D.A. Units—about 1922

heard with approximately equal volume, the set was eminently satisfactory.

Other means of regeneration were also developed, some replacing the tickler tuning with capacity feed-back or a resistance control. Another system put a variometer in the detector plate circuit so that it might be tuned to high impedance, causing the feedback to occur capacitatively between the elements of the tube.

None the less, the Armstrong regeneration patents were good, and while every manufacturer was eager to produce a regenerative set, not all were legally entitled to do so. A few resorted to clever methods of trying to beat the patent. One, to cite an example, incorporated a tickler coil in his set, but had it short-circuited with a heavy wire to which was attached a large red tag cautioning the purchaser not to remove the wire under any conditions, because it would make the set infringe. However, it added the information that the set would give much better reception if the wire were removed. Some purchasers, it is said, were able to wait until they got home before cutting the wire loose.

It was about that time—early in the 1920's—that the first all-wave sets became popular. These were honeycomb-coil outfits, a popular make being the Federal. This was a handsome 3-tube job in a cabinet approximately 2 feet long. The coils were on the outside, at the left, below them were 2 variable condenser controls and a series-parallel switch for putting the coils and condensers in series or parallel. The set had a variable grid-leak that varied with the weather, for it was merely a pencil mark between 2 contacts; a neat nickeled cap hid it from the public gaze when it was not being varied manually. If changing the switch did not extend the set's wavelength enough, one could always pull out the coils and plug in a fresh set—if one could afford them.

Another early type of multi-range set was the so-called Navy model. There really were some genuine Navy sets, left over from the World War, and these were readily distinguishable by their

PROGRESS OF RADIO
Receiving Equipment

dials—large metal affairs with movable pointers indicating the calibrations on their faces. The pointers were controlled by the same knobs which operated tap-switches, moving to inner or outer arcs on the dials as the switches were turned. Similar systems are used to indicate the band on present multi-range receivers.

Although sets which contained their detectors and amplifiers in a single cabinet were on sale, those which came as separate units persisted for some years. As late as 1926 such dual-unit sets were on sale, though they were being retailed through outlet stores at bargain prices.

All during this early period, receivers were battery-operated. The batteries were identified as filament or "A" batteries and plate or "B" batteries. Separate batteries to provide grid potential were not yet in use; such "C" batteries made their appearance during the middle 1920's and promptly won the public heart, not so much because they decreased distortion in the audio stages, but because they cut down plate current, making the "B" batteries last longer.

Costly as tubes were (some standard receiving tubes were $6.50 each; others, $9.00, and scarce to boot—during "the war" you paid a premium, besides, and liked it!) batteries were by no means the smallest expense which the set owner had to undergo.

Two types of filament or "A" batteries were in general use; standard No. 6 drycells for the 1.5-volt tubes of the WD11 type, 6-volt storage batteries for the type 201 tubes. Such storage batteries cost from $15.00 to $40.00, depending upon their quality and ampere-hour capacity. While many listeners used regular automobile batteries, which were somewhat cheaper, a good dealer could usually sell a special radio "A" battery instead, as it had thinner, more numerous plates. This was said to give it longer life and more even current flow.

While the larger ampere-hour-capacity batteries cost more than the lesser ones, they were an economy in the long run, as they required less frequent charging—and charging cost a dollar a shot

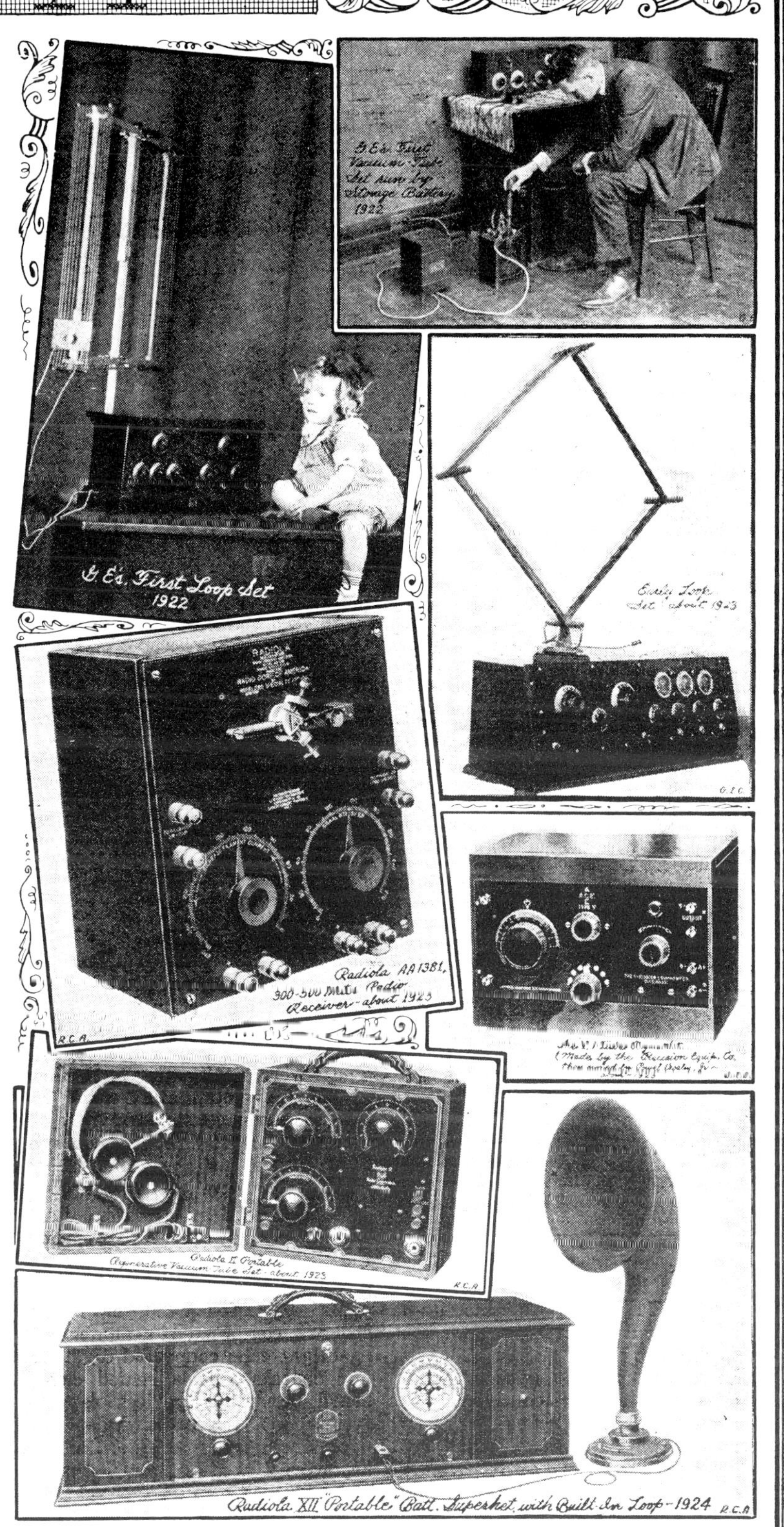

PROGRESS OF RADIO
Receiving Equipment

Radiola III A. Regenerative Det. and 2 A.F. (2nd Stage Push-Pull) about 1924 R.C.A.

Atwater Kent Model-12 ~ 1924 A.K.M.C.

Radiola Regenoflex. (Battery Compartments at left and right) about 1925 R.C.A.

6-Tube T.R.F. Low-Loss Resistance Coupled Set, with Electrad "Phasatrol" Oscillation Suppressors-about 1925 G.I.C.

Radiola-18 Early All-Electric Set (type 226-227 tubes) about 1927 R.C.A.

Philco's First Set 511 Series—1928 P.R.C.

Irvin (Noblitt-Sparks) Automatic (Pushbutton) Tuning Set 1938 N.S.A.

in most localities, with an additional 25c (per day) sometimes added for a rental battery for use during the charging period. Many listeners bought their own chargers at prices averaging around $18.50 and considered this an economy.

There were also trickle chargers, so called because of their low current output. These were connected through an ingenious switching system so that the battery was always on charge while the set was switched off. They were really the beginning of the "A" eliminator, which in its first form consisted merely of a battery across a reduced-voltage line, to act as a high-capacity electrolytic filter condenser.

Later the highly-filtered "A" eliminator came into being, but the "B" eliminator, as such, was made available first.

"B" batteries were, as many still are, composed of a number of flashlight cells connected together and contained in a sealed cardboard box. Their prices ranged from about $1.00 upward, depending principally upon size. Larger batteries lasted longer — save when some gyp manufacturer placed a small battery in a large box, filled the spaces with old newspaper or sawdust and sealed it up (or held a lamp or cheap meter on the terminals too long while demonstrating that the battery was up to par — which it probably was, before the start of the test!).

Eventually storage "B" batteries were produced. These

(Continued on page 614)

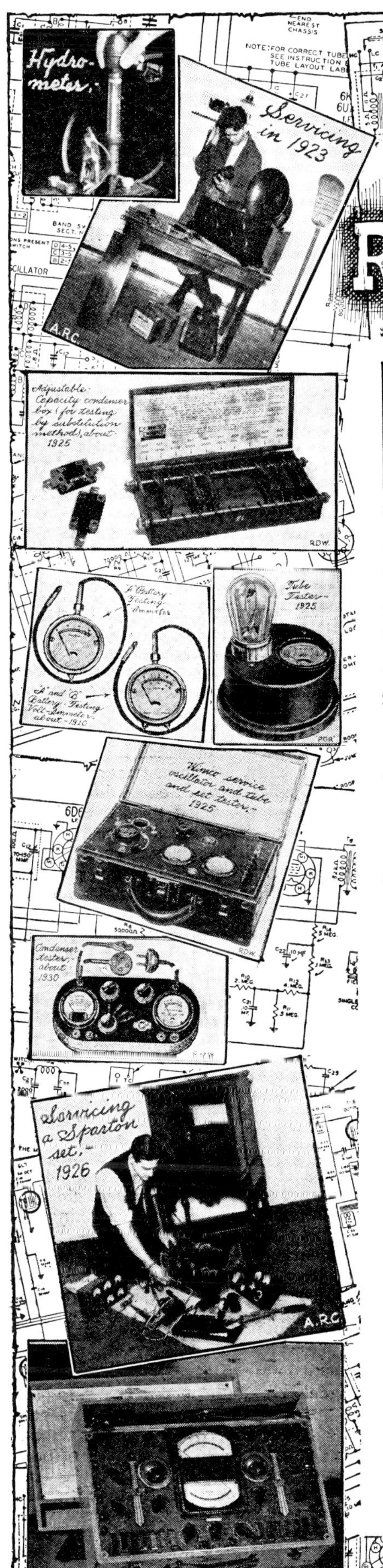

ORIGIN AND GROWTH OF RADIO SERVICING

IT is to be supposed that radio servicing started with the inception of "wireless"—back in the early Marconi days, since even the very earliest of apparatus which was in use must have required repairing at some time or other. In the strict sense of the word, however, the "Service Man" did not come into existence until 1923-'24, although radio technicians who performed repair work were an established unit with the United States' forces during the World War.

This condition was probably due to two factors: (1) receiving apparatus was so designed and constructed that it seldom, if ever, required servicing (especially in the case of crystal detector receivers) or, if it did, it was either a case of replacing tubes or batteries (in the earliest regenerative receivers) which was so simple a job that either storekeeper or consumer did it; (2) real technical equipment, such as transmitters (spark or tube), was installed and maintained by factory technicians who were usually the engineering representatives of the manufacturers.

SERVICING IN 1923-'24

It wasn't really until 1923-'24 that the commercial broadcast receiver began to take on a technical aspect, which accounts for shopkeepers or their "helpers" ceasing to replace the set's batteries (with the aid of a 0-50 volt pocket voltmeter and a hydrometer for checking storage batteries), and a sudden demand created for men with a knowledge of radio to perform whatever repairs were found necessary. From simple regenerative sets with 1- or 2-stage audio amplifiers manufacturers suddenly began to make receivers with reflex circuits, stabilized T.R.F. stages, superheterodynes, etc.—although all these receivers were still battery operated. A good many of these sets were of the custom-built variety, especially those with superheterodyne circuits. However, set manufacturers began at that time to use the complex circuits, as well as compact and elaborate construction, all of which necessitated skilled or trained men to service them.

The De Forest Radio Tel. and Tel. Co., one of the earliest pioneers in the field, was manufacturing the famous D-7 all self-contained loop receiver in 1923, which used 4 tubes, or stages, 3 of which were reflexed so that 7 stages of amplification were obtained, plus a crystal detector. In 1924, this company followed up this trend in design with another 4-tube reflex, the D 12, a really handsome leatherette semi-console cabinet job with inner compartments for the "A" and "B" batteries; speaker within cabinet with adjustable speaker unit; folding-type loop with calibrated compass plate on top cover (through which the loop would go into a jack receptacle) with plug tip end so that it could be removed and an aerial and ground substituted; 2-gang condenser to tune 2 circuits, and many other novel refinements.

THE RCA "PORTABLE"(!)

That same year (1924) RCA came out with their "portable" superheterodyne using 199-type tubes, with all batteries and loop self-contained within the case,—a rather gigantic affair (over 36 ins. in length) and very heavy but intended also for home use.

Within a short time of their introduction both the De Forest and RCA sets required servicing; in the former the A.F. transformers would burn out and the speaker unit developed mechanical troubles due to the adjusting feature. In the RCA sets the well known *catacombs* were giving trouble, although as a rule most of the troubles were due to the 199's which always were extremely fragile tubes. At any rate, this servicing work was beyond the scope of the dealer and a demand for Service Men was created. Strangely, despite the great number of constructors that

(Continued on page 651)

A CHRONOLOGICAL

Since the dawn of civilization man has ever tried to improve the art of communication. Delving into the history of mankind, we find the smokefire and sema-

STATIC ELECTRICITY - 600 B.C.

J. T. BERNSLEY

ANY history would be considered incomplete if it did not include all events from the beginning. Hence, since radio is based on electricity, our chronology must begin with the ancient days, with the "discovery" of the peculiar, spark producing properties of *amber* when this material was rubbed on a piece of cloth or fur. Similarly, knowledge of magnetic attraction in *lodestone*, which also dates back to the days of the ancients, was an epochal event; inasmuch as both of these "accidents" were responsible for the later discovery and refinement of electrical laws and principles, which served as the foundation upon which "wireless" was built.

While no exact dates are available, the earliest histories mention the phenomena surrounding amber and the lodestone, as far back as 600 B.C. In that era, it is chronicled, Thales discovered the mysterious sparks which resulted when rubbing the mineral amber, and which we now know to be "static electricity." Since the Greek word for amber is "electrum," when the experiment was repeated many centuries later, so that more might be known regarding its cause and effect, it served as the root for a new word—*electricity*.

During the long interim, strange and fantastic superstitions were conceived concerning static electricity and magnetism. The philosophers of the early days theorized quite a bit, but did very little experimenting; and what was known was handed down from generation to generation with elaboration and no attempt at justification by, at least, trial-and-error experiments. Consequently, we find such fallacies, based more on hearsay, as "garlic odor destroys the magnetic potency of the lodestone or the compass." This myth lasted through the early centuries clear up to 1544, when the famous treatise on Physics by Philip Melanchthon included mention of it. After that time, numerous controversies sprang up, pro and con, until 1646 when it received its death blow from Sir Thomas Browne. This astute physician-scientist refused to take anybody's word for it, and actually performed experiments with magnetized iron and "garlick juice"—thus definitely disproving the centuries-old superstition.

Similarly, numerous theories were created concerning amber and diamonds; among them the most notable being that iron rubbed with a diamond became a magnet, and that diamonds when rubbed would attract bits of paper and particles of dust. Another English scientist, Dr. William Gilbert, outraged at what he termed "chattering of barbers," undertook to disprove these theories by actual experiment. To his discomfiture, he found that rubbing diamonds *did* cause them to attract bits of paper; but he discovered also that practically everything he rubbed, except metals, became thus "electrified."

This led Gilbert to compile a huge list of materials which could be "electrified," including such items as "true jewels and paste imitations, sulphur, sealing wax, rock salt, alum, resin," etc. It was this gentleman who gave the name "electric" to this effect, and later on, in 1675, Robert Boyle, in his "Mechanical Production of Electricity," coined or derived the word "electricity" from it. More important, however, is Gilbert's invention of the "electroscope," which he probably used to test the various materials enumerated in his lengthy compilation. Also, he too set a precedent by conducting actual experiments before publishing scientific information, refusing to accept hearsay as fact.

With the beginning of electricity, came a new era in scientific research; from then on, scientists have resorted to trial and tests to confirm their theories. Back into the darkness were dispelled the mysteries and superstitions of the ancients. Electricity was born, and new fields were open to conquest. Perhaps that is why some chronologers refer to Gilbert as the "Father of Electricity." At any rate, since this period marks the inception of scientific electricity, because of the gradual elimination of rumors, guesswork and superstition, our chronology begins with the date of 1600. Without question, scientific work of importance in this field was reported earlier than this period, and this fact is now acknowledged to avoid confusion, discussion and unnecessary arguments.

1600 Dr. William Gilbert, physician to Queen Elizabeth, and scientist. Invented the "electroscope," consisting of a straw which was pivoted like a compass needle, and which indicated the approach of a charged body. Disproved many myths, and compiled a list of materials which could be electrified by rubbing. Coined the word "Electric", from the Greek root for amber—"Electrum." Conceived the earth as a huge magnet, with magnetic poles and a field of magnetic force about it—thus laying a positive foundation for many scientific discoveries to come later.

1646 Sir Thomas Browne, English physician and author. Performed many experiments with the lodestone and magnetism, refuting many idle superstitions by actual trial. Actually tried to make the first "wireless" by employing two compasses with the alphabet written about them (although credit for the idea must be given to a predecessor, one John Baptista Porta). He imagined that, if the two needles were magnetized together, then separated, the turning of one to indicate some letter of the alphabet would cause the indicator on the second dial to move to a similar position; thus envisioning a means of communication without any intervening medium. The second compass indicator, however, did not budge from its North-pointing position, remaining as he said "like the pillars of Hercules." Nevertheless, the thought of communicating between persons over a distance, hitherto more or less a figment of the imagination that inspired very few people in the centuries before, became an intriguing thought in the minds of many scientists of that time.

1672 Otto von Guericke, German burgomaster of Magdeburg. Famous for the "Magdeburg experiment" with which he proved atmospheric pressure, and entertained king and princes. Built an "electric" generating device of a globe of sulphur mounted on an axle and turned by a crank. The globe was rubbed by the dry palm of the hand, as it rotated and, after some little friction, the globe was sufficiently electrified to attract particles. This machine was, of course, a generator only of *static* electricity, not current electricity which we now use. While experimenting with this device, he discovered that the particles after they had been attracted, would in a

HISTORY OF RADIO

phore means of signaling an ancient device for communication. Radio is the result of a series of experiments in communication which we here chronicle.

SHEPHERD DISCOVERS MAGNETISM — PRIOR TO 1000 B.C.

short while be repelled. At this time we know that the particles assumed a like charge to that of the sulphur ball and, when this condition occurred, the particles were repelled because "like charges repel each other." But the poor burgomaster was laying the foundation for our knowledge by chasing a repelled feather around the room with a heavy globe of sulphur in his arms. In his pursuit of the feather, he noticed that the feather was repelled by a lit candle on the table, and then "flew back to the sulphur globe as a sort of a guard." Von Guericke attributed human attributes to a feather when, in reality, he had observed the fact of electronic emission without knowing it; nearly three more centuries elapsed before anyone knew more about this phenomenon. What really had happened to the feather was that its charge was dissipated and then changed by the electron stream from the candle and, consequently, was now attracted to the globe which had repelled it before. His experiments resulted in further discoveries but, as with all pioneering work performed with little or no background, satisfactory explanations for some of the phenomena he encountered were not available. He heard and saw the tiny discharges which resulted when he generated static electricity with his globe, but he didn't associate it with lightning and thunder. That discovery was to wait until Franklin's experiment with the kite.

1729 STEPHEN GRAY, Englishman. By experimenting with charged bodies, Gray discovered the effect and the differences of what we now know to be conductors and insulators, as regards conveying or transmitting charged impulses. About this same time, DUFAY, a Frenchman, conducted similar experiments but along more elaborate lines. He, evidently, was versed in Gray's accomplishments because his work seemed to be in the nature of proving or disproving Gray's discoveries. In the course of his experiments he found that metal wires or wet objects were the best conductors, though the most difficult to electrify, while those easiest to electrify were the best supporters or insulators of the charged impulses. In fact, he built a line, a quarter-mile long, which consisted of a wet thread held up on glass tubes and determined that it was an excellent means of conveying a charged impulse from one end to the other. This was probably the first transmission or electric line, and consequently an important discovery.

1745 PIETER VAN MUSSCHENBROEK, of Leyden, Holland. Invented the Leyden jar, after discovering it in an accidental but most interesting manner. It must be remembered that the scientists of this period were still playing around with friction apparatus, since no other means for generating electricity had been discovered. Musschenbroek had the thought that electricity could be bottled or, rather, confined within a bottle so that it could be used at some later time. Whether the idea was original with him is hard to determine, since histories vary. At any rate, the idea was that, if water were placed within a bottle and then charged by means of a frictional-electric producing machine, the charge would remain in the corked or stoppered bottle because glass is a good insulating material. Fate took a hand the day Musschenbroek was conducting the experiment. He was turning the crank of the electric-producing machine, while his assistant, Cunaeus, was holding the jar with one hand and with the other trying to draw off sparks from a gun barrel. The circuit consisted of the gun barrel connected to the friction machine and also to a brass wire which entered the jar, partly filled with water. Had Cunaeus placed the jar on a table, nothing would have happened, and the condenser might not have been heard of today. As it was, his hand formed one plate, the liquid in the jar the other plate and, while Musschenbroek cranked the machine, the improvised condenser eventually became charged up—and then Cunaeus must have thought the world had come to an end! The tremendous spark which resulted caused the entire charge to pass through his body—and the records have it that Cunaeus was incapacitated for two full days. Another scientist of that period, Nollet by name, heard of the experiment and, unwilling to be a subject for experiment himself, got together approximately two hundred soldiers, had them all join hands in a large circle, and then, in much the same manner as Musschenbroek and Cunaeus had done, sent a severe charge through them. The fact that they all jumped instantly and strenuously pleased him immensely, and gave him much to marvel at. Naturally both Musschenbroek and Nollet tried to figure out what had caused the effect, and it wasn't for some time that a definite conclusion was arrived at. They found that, when they placed the jar of water on a table, it would refuse to be electrified (since the other plate of the condenser was lacking) and that, only when the hand was placed around the jar, could the phenomenon be repeated. But volunteers for the experiment were probably lacking; so eventually it was discovered that placing the jar over a metal plate seemed to do as well. Later on, an outside tinfoil covering was substituted, with improved results, and for many years this was the actual construction of Leyden jars the granddaddy of all condensers.

1751 BENJAMIN FRANKLIN, American statesman, philosopher and, last but by no means least, scientist. Practically everyone is familiar with Franklin's kite and lightning experiment—but perhaps too familiar with this phase of his work and not so well versed in his other scientific endeavors. Some of his deductions have played an important role in the development of electricity since he employed the same methodical precision and calm logic which made him famous as a statesman and philosopher. Franklin established the law of conservation of the electric charge; that there are a Positive and Negative kind of electricity; that lightning and thunder are related to the crashings and sparks obtained when electrically-charged bodies became discharged. He invented the lightning rod, to prevent the great damage done to property by lightning, and sent the suggestion to the Royal Society in London—but was ridiculed for it. His theories led to his followers' discovery that air may be substituted as the dielectric in place of glass in the construction of a Leyden jar, as well as that "like charges repel and unlike charges attract"—which is now axiomatic.

1780 ALOYSIUS GALVANI, Italian professor of anatomy. Up to his time, only two means for obtaining electricity were known; one by means of the frictional machine, the other from the clouds, as discovered by Franklin. Galvani (by accident, it is reported) noticed that an electrical charge applied to a dead frog's nerve would make it kick and struggle as if it were very much alive. Continuing his experiments along this line, he found that a number of frogs he had prepared and suspended on his balcony would respond to lightning flashes in similar manner and that, even before a storm, if a frog's legs happened to touch the iron part of the balcony, the twitching muscular movement would occur. Later on, he determined that any two metals joined together, so that one touched a leg muscle and the other a leg nerve, would cause the muscular twitching. Galvani then reasoned that the muscle was akin to a Leyden jar, and that the electricity was a fluid which made a circuit from the muscle to the nerve, then through the metallic conductors back to the muscle again. He called the "fluid" animal electricity; but true *galvanic* electricity, as caused by two dissimilar metals in contact, was not recognized by Galvani who theorized that the electricity originated in the frog's leg.

1790 ALESSANDRO VOLTA, Italian professor. Shortly after Galvani's experiments, Volta devised what we now know as the "voltaic pile," consisting of a pile of alternate zinc and copper discs (each pair of discs being separated by a moistened pasteboard disc and termed a "couple"); so that, by using quite an aggregation or large pile of discs, a distinct shock was obtained when the finger tips were placed on each end of the pile. The disadvantage of this arrangement was that, when the pasteboard discs dried out, the voltage diminished. Consequently Volta devised copper and zinc strips, joined at the ends and placed in separate jars containing a weak acid solution. Now we have the first real *battery*—a unit destined to be of great help to future inventors and scientists in their explorations into the realm of electricity. In honor of this discovery, Volta's name was immortalized when, later on, the *volt* was the name given to the unit of electrical force.

1800 NICHOLSON AND CARLISLE, English experimenters. Set up a voltaic pile and showed that water could be decomposed into its elements, hydrogen and oxygen, by passing an electric current through it. Known now as the electrolysis of water.

1820 HANS CHRISTIAN OERSTED, Dane. Professor at Copenhagen. For thirteen years Professor Oersted had experimented with electricity and its effect on a compass needle, having read in Benj. Franklin's reports that there was some effect and relation between the two. While lecturing to a class, Oersted had his attention called to the wavering of a compass needle, whenever a switch was thrown which connected to a voltaic pile. After the classroom students had departed, he investigated the phenomenon—finally ascertaining that, when the compass needle was placed *along* the wire, there was a deflection, with the compass needle coming to a stationary position when it was *across* the wire. When the compass was placed *above* the wire, the needle turned one way, when placed *under* the wire it turned the other way. This was the basis for determining magnetic lines of force, and without doubt the foundation for measuring or indicating electrical instruments. In this same year, the chronicles have it, one week after Oersted made the aforementioned discovery, Andre Marie Ampere, French scientist, made the important discovery that two parallel wires carrying an electric current but free to move, attract each other if the currents travel in the same direction, and repel each other if they travel in opposite directions. Also, he determined not only that a wire carrying an electric current would attract a magnetized needle, but that the needle would also attract the wire. Today we find the unit of current, the *ampere*, named in his honor.

1826 GEORGE SIMON OHM, Bavaria. His outstanding accomplishment is the law which now bears his name: "A current flowing in any closed circuit is proportional to the force or voltage and inversely proportional to the resistance of the wire." Today we express *Ohm's Law* simply by mathematical means, viz., $I = E/R$.

1831 JOSEPH HENRY, American physicist, improved the electromagnet (developed by Arago in 1820) by using silk-covered wire, which allowed the use of many layers of turns. First to employ insulated wire, which permitted him to make coil-magnets large enough to lift several pounds. The unit of inductance, the *henry*, is named after him.

1832-1837 SAMUEL F. B. MORSE, American artist, created the electric telegraph system and conceived a "code" which permitted transmission and reception of messages. This *Morse Code*—still used in wired telegraphy—was soon adopted for use in the earlier transmission and reception of wireless messages.

1825-1867 MICHAEL FARADAY, English. Since it is very difficult to assign accurately the various dates for Faraday's numerous inventions and discoveries, we herewith list the period of his activity. In 1824 Faraday became a Fellow of the Royal Institute, but his fame as a scientist had preceded this date. He died in 1867; and in the interim his discoveries were the most complete, numerous and productive of any contemporary scientist's. They deal with every phase of the sciences, physics, chemistry, mechanics, electrochemistry, and electricity. His first explorations in the field of electricity resulted in the basic principle of the electric motor. Faraday reasoned that, if an electric current in a wire causes a magnetized needle to rotate, then a magnet should cause a wire carrying current to do likewise. He proved his reasoning by suspending a conductor, so that it could rotate between magnetic poles. He formulated the laws of magnetic induction, which finally led him to invent the first electric generator; as a matter of fact, he built many models, each time improving them. He invented the induction coil which was later improved by Ruhmkorff, a Frenchman; and also the transformer, which operated from alternating current and, consequently, did not need the

HISTORY OF RADIO

interrupter device for starting and stopping the current. In fact, Faraday discovered alternating current; and the experiment now shown in high schools, for producing electricity by plunging a bar magnet into a coil of wire, was conceived by this most brilliant of all inventors. He made a study of condensers, discovered different dielectrics that may be employed, and analyzed the relative merits of each—finally tabulating this data so that today we have the "dielectric constant" for each insulating material and can be guided accordingly. He coined many electrical terms now in use.

1865 JAMES CLERK MAXWELL, Scottish. Elaborated mathematically what is known as the "electromagnetic theory of light", although the thought was conceived by Faraday. This theory says that light, electric waves and magnetic waves, of varying frequency, travel in the same medium, namely —ether. Since ether permeates all matter, a current may exist in and about a conductor, but is essentially guided by it.

1865 DR. MAHLON LOOMIS, American dentist. The inventions of the previous years in the field of electricity had brought about the electric generator, batteries, the telegraph, arc lights, a trans-Atlantic cable and many other devices which were a great boon to humanity. But man is always continuously striving to improve as well as explore, and so we find Loomis, a Washington dentist, conducting experiments and applying for a patent on a method for transmitting and receiving messages whereby the earth's atmosphere is used as one conductor. Strangely, he not only wanted to send messages as aforementioned, but also to do away with batteries or generators, since he was acquainted with the fact that the atmosphere is continuously charged with electricity. Operating on the theory that the higher the level, the greater the charge would be, Loomis sent up kites 18 miles apart, from two high mountains in West Virginia. The kites, covered with large squares of copper screen or gauze, were connected to the ground by strings within which fine copper wires were enclosed. The wire from each kite string was connected to one side of a galvanometer, the other side of which Loomis held in readiness, so that he could establish a connection to a coil buried in the earth. The receiving station connection, between meter and earth coil, was always closed; and, whenever the circuit was closed at the transmitting end—wonder of all wonders—the galvanometer at the receiving station actually dipped! This and other numerous similar tests were conducted in the presence of reputable witnesses; and Loomis almost got an appropriation of $50,000 from Congress, to develop his invention.

1875 IN 1875 the microphone (or magnetic transducer, which functions optionally as an earphone) was invented by Alexander Graham Bell. About 2 years later D. E. Hughes invented the carbon microphone.

1879 DAVID EDWARD HUGHES, English. Discovered an arrangement which consisted of a stick of wood covered with powdered copper; when placed in an electrical circuit the copper particles would cohere when a spark was made.

1885 SIR WM. H. PREECE and A. W. HEAVISIDE, Englishmen. These two gentlemen sent signals to each other over a distance of 1,000 yards. The means employed consisted of two telegraph lines paralleling each other, with a telephone receiver in the receiving side. The telegraph signals could be clearly heard in the phone receiver, without actual connection between the two, due to what is known as *induction* or, in common telephone parlance, "cross-talk".

1887-1888 HEINRICH HERTZ, German. It will be noted that the dedication of this issue of *Radio-Craft* to old-time radio begins with this date since, in reality, Hertz' experiments paved the way for Marconi's work in this field. Some prefer to call Hertz the "Father of Radio"; and that he deserves more than ordinary recognition, for his work in this field is indicated by the fact that radio waves are commonly referred to as "Hertzian waves". Hertz first became intrigued with this problem when he studied Maxwell's theories concerning light, magnetism and electrical waves. In an attempt to gain further data on this theory, Hertz actually set up the first spark transmitter and receiver. The transmitter consisted of a Leyden jar and a coil of wire, the ends of which were left open so that a small gap was formed. For the receiver he employed a similar coil, with gap arrangement, located in the opposite end of the room. When the Leyden jar was charged sufficiently, it discharged through the gap in the wire coil; and the oscillating waves thus generated were launched into the ether of the room, and swept across the receiving coil causing sparks to fly across the gap in the receiving coil. Hertz measured the velocity of these waves and found that they were the same as that of light, 186,000 miles per second; also measured their length and, subsequently, substantiated Maxwell's theories.

1892 EDOUARD BRANLY, French. Inventor of the coherer, which was later destined to play so large a part in the practical reception of wireless waves by Marconi. The coherer was not named as such until later, nor was it basically conceived by Branly, since Hughes had employed a similar device as mentioned previously. Branly, however, made the device as Marconi was to use it, consisting of a tube containing loose zinc and silver filings, and plugs to make contact to each end. Since the filings would cohere (stick together) after the first spark was received, a means of separating them for the next signal was necessary. Popoff (Russian) conceived the idea of employing the vibrator and hammer of an ordinary electric bell in the circuit of the coherer so that, almost the instant the filings cohered, the hammer would strike the tube and cause them to "decohere".

1893 NIKOLA TESLA, Serbian. Suggested a means of wireless communication which utilized the earth as a conductor and created stationary electrical waves on it. Invented the Tesla coil, which, in effect, created high-frequency oscillations of a broad nature (hence was in reality a broad wireless transmitter) but, since he made no effort to detect

(Continued on page 642)

Reminiscences

Pause for a few minutes and pass back over the years with a few of the famous old-timers who made radio history. Radio-Craft *is proud and deeply honored to record for posterity this glowing record of hard-fought achievement.*

Dr. LEE de FOREST—

Lee de Forest Laboratories

RENOWNED for his discovery of the Audion,—the first 3-element tube! One of the first of American radio pioneers and manufacturers. Designed and sold many early radio stations, including radiophone types, some of them to the U.S. Navy. His pioneering work in the development of "film" talkies and other researches are also well known to many.

FIFTY years of Radio! My memories don't go back quite that far. Thank heaven, because otherwise I fear my future efforts would be too briefly curtailed—and it's a great game, this Radio, which I don't want to abandon, not for many years yet to come.

No, 37 years is all I can lay claim to—first in wireless, then in Radio experimentation, invention, and manufacturing.

My recollections of the earliest beginnings are still vivid—my first tests from the flagpole of Armour Institute to the Lakato Hotel—a half-mile; and the ecstasy of jubilation I thrilled to when those first purring signals from my single headphone entered my ear. No triumph since that day has ever excited me as did that first, primitive success.

Next the tests to the tower of the Auditorium Hotel, then to a yacht on Lake Michigan, actually sailing the horizon, and I heard the signals—"amazing, incredible"!

Then came the battles in Manhattan, to persuade men of wealth that wireless telegraphy was (1) possible, (2) profitable. None, old J. P. Morgan, Geo. Westinghouse included, could be persuaded that wireless possessed potential earning powers. Believe it or not!

Came the 1903 International Yacht Races. "Sir Thomas Lipton installs de Forest Wireless on his yacht *Erin*"—The A.P. transmitted thousands of words to my Coney Island station, and American business began slowly to realize that perhaps "we had something there". The demonstrations at the St. Louis World Exposition convinced the public—commercial wireless messages to Springfield, then Chicago—300 miles overland, an unheard-of world record!

And then the U.S. Navy began to install "modern" American wireless on battleship and cruiser. T. R.'s famous trip to the Canal on the *West Virginia*, in constant touch with shore, even exchanging messages with Ohio's Governor Herrick, from mid-Gulf to my Cleveland station; then the Navy's 5 world-beater, 25-kw. (!) stations in the Caribbean and Porto Rico, resulting in rapid installations on most of our Atlantic coast marine.

But O.M. static in 1906 caused collapse of financiers' over-ambitious plans for overland networks; the wireless stock bubbles burst, and when the smoke subsided American wireless was owned and controlled by London.

So I began over, this time with "Radio", for words and music, not merely dots and dashes. Solely salvaged from that wireless wreck was a small glass bulb, a "wireless detector", fortunately (for me) deemed commercially worthless in comparison with the slightly less sensitive, more rugged crystals.

Using a small carbon arc in alcohol flame as my undamped-wave generator, with double or triple pancake tuner and the grid Audion detector, I was soon broadcasting phonograph music for 10 miles from a small laboratory, atop the old Parker Building—4th Avenue at 19th Street. There the 3-electrode amplifier for telephone currents was invented, destined later to twine voice networks around the globe.

November, 1907, saw Bob Evans's Fleet start around the world, every vessel carrying a radio telephone. The uncertainties of performance of the arc-transmitter, with husky carbon mike in the antenna-earth lead, rendered these early broadcast efforts abortive.

So again I retired to a laboratory for new studies, further discoveries; this time in far-off Palo Alto. There in the summer of 1912, while constructing the original Cascade Audion amplifier, for later demonstration to the Bell engineers in New York, the first "feed-back", or regenerative circuit was accidentally discovered.

One year later and Charley Logwood and I were telephoning across my Highbridge Laboratory using a small audion bulb for transmitter, another as detector, a third for amplifier to headphones. Thence grew quickly "oscillion" tubes, of 25, 50, 100 watts—grid modulated; and actual broadcasting—with Election Returns in November, 1916—4 years before WWJ and KDKA, and nightly or thrice-weekly programs, until the War stopped everything.

The post-war growth, from 1919 to today, is history familiar to most of your readers. But those earlier struggles, disappointments, defeats, ultimate triumphs—their memories are cherished by but a few veterans and pioneers, in numbers steadily, sadly diminishing—annals which now make pleasant recollection.

DAVID SARNOFF—

Pres., Radio Corporation of America

WHO, in the radio industry, has not at some time heard of David Sarnoff and the "Horatio Alger"-like story of his success? From lowly wireless operator in 1907, to president of the Radio Corporation of America; that interim has been full of successful incidents that now make this man an outstanding figure in radio, and in the business world.

IN 1888 an American author, Edward Bellamy, published a novel entitled "Looking Backward." It became a best-seller, was published in many editions and translated into many languages. The author pictured a Utopian world of the year 2000, where, among other miraculous advantages, one might summon beautiful music into one's home merely by turning an electric switch.

Of Old-Timers

Albeit many of these lines are whimsical, none-the-less, reading between them we see years of disappointment and disillusion that only the indomitable spirit of the pioneer was able to surmount. Radio's spirit of yesteryear carries on!

In 1888 Heinrich Hertz was probably too busy experimenting with an electric spark and a loop of wire to read "Looking Backward." In any case it may never have occurred to him that there could be any association between his experiments and the early fulfillment of Bellamy's dream. Nor could he have believed that the work he was doing would make possible instant communication between all nations; would promote human safety on land, at sea, and in the air; and would help bring not only music but news, education, drama, and comedy into the home.

Although radio as we know it today has been a half-century in the making, its most spectacular progress has occurred only within the past 15 years. As radio has gained momentum its forward speed has been constantly accelerated.

Prominent among the accomplishments of the recent past are the automatic SOS radio alarm, which insures reception of distress calls at sea; the improved application of radio technique to recorded music, and to motion picture films; the widespread development of automatic tuning; the development of shortwave technique to permit broadcast programs of all nations to be heard around the world.

Radio, the pioneer, continues to scan far horizons. Television is steadily progressing toward the ultimate goal of a useful public service. Recent outstanding television advances have included the development of more sensitive iconoscopes, and of larger kinescopes, presenting black-and-white instead of tinted pictures; the projection of television pictures onto a 3 x 4 ft. screen; the successful transition from 343- to 441-line scanning; and the development of a mobile truck-unit for outside program pick-ups.

There is only one thing certain about the future of radio, and that is the certainty of fresh surprises. If, in 1938, another Bellamy should write another "Looking Backward" his imagination would be unequal to the task of predicting the ultimate results of the simple findings of Hertz and their application by Marconi. They are beyond human calculation. *Radio has only started.*

Dr. FRANK CONRAD—

Assistant Chief Engineer, Westinghouse Elec. & Mfg. Co.

MODERN broadcasting owes its inception to Frank Conrad's earlier activities as an amateur. In 1919, while employed as an engineer with Westinghouse in Pittsburgh, Conrad began to send out programs from his station installed in his garage. Westinghouse became interested in the publicity that resulted, and KDKA (the first broadcaster) was the result.

AMONG the many incidents associated with radio broadcasting I recall, none remains clearer in mind than a demonstration of short-wave transmission between our station at East Pittsburgh, Pa., and London, England, made before delegates to an International Radio Conference held in 1922.

Misconception as to the efficacy of the short-wave band for distance transmission was wide-spread at the time due to the well-known skip-distance phenomenon of ultra-high-frequency, whereby they tended to fade out a short distance from the station only to come in strong again at points far removed.

For several years, those with whom I associated had been conducting tests with short waves and were familiar with the characteristics of such transmission, finally concluding that it would be an excellent medium for trans-oceanic operations.

In 1922, I carried a short-wave set with me to London, keeping in touch with our Pittsburgh radio station while on board ship.

At the conference, the question of short waves was discussed and some questions were raised as to their ever being of any value in radio.

One night, we invited a number of delegates to my hotel room and using the short-wave receiver, with a brief length of wire tossed out the window, listened to Pittsburgh. By cable, I had arranged for a complete program to be sent from our Station W8XK. The announcer in Pittsburgh, probably thinking to bring us up to date with the news, read not only the entire front page of a Pittsburgh newspaper but also several other columns inside.

The listening radio delegates heard not only their first trans-Atlantic short-wave transmission but also the greatest number of words sent over the ocean by radio up to that time.

The incident was widely discussed at the conference and the short-wave band given due importance in the consideration of future radio requirements. Now, of course, the short-wave band is used widely in radio transmissions and will continue to be the part of the radio spectrum from which new channels for radio use will be drawn.

POWEL CROSLEY, Jr.—

Pres., The Crosley Radio Corp.

PIONEER manufacturer of radio receivers and parts, and now president of one of the largest radio set companies and largest radio broadcast station in the world. His vision and faith in the future of radio are well illustrated by the progress and success of his company, since the days when he first made the "Pup" 1-tube set, and book-type condenser.

YOU ask me to reminisce a bit about radio. So here goes.

Perhaps you have heard how my son, then 9 years old, came to me back in 1920, wanting me to buy him a wireless outfit. I did—a little crystal detector affair. That is how the whole thing started. Now, son is Vice-President in direct charge of our radio division. In the meantime, we have built several million radio sets.

Then again I look back to our first broadcasting station

Reminiscences

—a little 20-watter—upstairs in a room housing a billiard table that I used as a laboratory in the early days of radio. The call letters were 8CR. It was an amateur radio television transmitter. At odd times I used to put phonograph records on. One day with the wind blowing in the right direction I received an acknowledgement card from Lima, Ohio, 100 miles away. Marvelous!

I remember when a radio club in Cincinnati got after us pretty hard for using super-power—50 watts. There have been similar repercussions down through the years in pioneering 5,000, 50,000, and 500,000 watts. But I still say that a lot of people have been able to enjoy better radio reception in remote parts of the country as a result of such pioneering.

I look back upon the day when we first carried on 2-way communication with WLW hooked-in with an airplane in the sky, directing its course to the location of our transmitter then at Harrison, Ohio. I remember the chap who came into my office with the idea of buying a "Jenny" plane. He wanted to paint our trade-name on the lower side of the wings. I thought I had a better idea—load the ship with a receiving set and powerful loudspeaker; transmit by radio to the plane and let it talk back to the ground. So we tried it and it worked. That was a long time ago as time goes in radio.

I could go on enumerating hundreds of little incidents throughout the past 16 or 17 years, but one series of incidents that stands out very pleasantly in my mind has been the meetings from time to time with you, Hugo Gernsback. While our paths have not crossed frequently recently, I read your stuff regularly and admire the inventive vision of your brain. How many things have come true that you have predicted!

Dr. ALFRED N. GOLDSMITH—

Consulting Industrial Engineer

PAST-PRESIDENT of the I.R.E. and S.M.P.E., Dr. Alfred N. Goldsmith still edits the Proceedings *of the former body and is on the governing board of the latter. Some years ago he taught electrical engineering at the College of the City of New York, and more recently was Vice-President and General Engineer of RCA. He is now a consulting engineer.*

ONE-THIRD of a century is not much in the life of nations but it is a long time in the history of radio. At the beginning of the 20th century the "ether" was practically undisturbed. Except for lightning storms and occasional X-ray outfits all was quiet on the electrical front.

The radio amateur or professional of those days needed a lot of patience. To build apparatus that would work, he required infinite tenacity together with outstanding detective ability. People who knew anything about radio (or "wireless" as it was then called) kept their dark secrets to themselves. Accordingly it was necessary to extract the facts by force or guile.

I can recall building a portable coherer receiving outfit in those days. It was carried to the homes of my friends, and aerial wires were lowered out of windows or thrown across roofs. Then began a long vigil in each case. After anything from 10 minutes to several hours, the radio telegraph signals from some ship in the harbor would gladden expectant ears. The reaction of my hosts varied, but was generally either amazement, incredulity, or else a slight doubt as to the genuineness of the actual reception. Messages were few and far between and tuning, as we now know it, played little part in those days of "high-decrement waves".

Later, radio telephony began to be of interest. One of my friends, an eminent radio engineer since deceased, had a small arc telephone transmitter installed in lower Manhattan. He visited my laboratory to pick up the signals on a simple crystal receiving set, vacuum-tube amplifiers being unavailable in those days. The tests were carried out in the silence of night in a quiet neighborhood. The headset telephones were closely clamped to the ears of the expectant listener. Then the telephone cords were held away from the table and equipment by his companion, because *even the slight scratching sound of a telephone cord rubbing against any obstacle would have been sufficient to drown out the excessively feeble telephone signals* which could then be received over even a few miles distance!! Such was radio telephone reception in the days before power output tubes and efficient loudspeakers.

Less than 10 years later it was possible to put many hundreds of watts of speech-modulated energy on the air. A radio telephone transmitter which I had in operation more than 20 years ago, working around 1,700 meters, was received in the mid-West on midnight tests by an eminent university professor who was also a radio expert. This gentleman later became one of the leaders in the field and a high officer in the communication service of the United States. At midnight he would send me, after each test, a wire telegram telling me of the results— after which I would go to the local telegraph office near my laboratory and wait for the telegram to arrive (there having been no messenger boys available at 1 o'clock in the morning in outlying districts in those days).

Altogether, radio was a rough and ready field one-third of a century ago but it was full of interest, new things happened frequently, and all radio men were enthusiasts. Indeed they had to be enthusiasts to stay in a field which most sober business people regarded as related to mesmerism, hypnotism, or some other exotic "ism". Now radio is respectable and occasionally even stolid. But fortunately the new arts of radio centering about the ultra-high-frequency waves, facsimile, and television will provide both the headaches and the stimulations which will make radio interesting and worth-while as a career in the future.

ELLERY W. STONE—

Vice-President, All America Cables, Inc.

PROMINENT radio engineer and consultant in the old days when radio was booming. Some of the Pacific coast radio amateurs and wireless "ops" will remember him as their inspector (San Francisco district) between 1914 and 1917. His book, "Elements of Radio Telegraphy", served as the text for many student operators and amateurs of those early days.

I REMEMBER when *Modern Electrics* was my engineering guide and its companion enterprise—the Electro Importing Company, generally known as the E. I. Co. to

Of Old-Timers

the "aficionados"—was my only source of radio equipment other than that which *Modern Electrics* taught me to build.

As a boy in high school, radio was my first hobby. After 30 years of it, it is still as fascinating. In the old days, the hobby end of it was as a "ham" operator. Today, the telegraph end of it is my daily job and the hobby end of my radio interest has drifted over into high-fidelity reproduction of music—radio and phonograph. But once a radio man—always a radio man.

So I remember when my first detector was a carbon and needle. Then I went to a crystal and graduated through carborundum, silicon and galena to reach finally in 1913 the dizzy heights of the de Forest audion—the low-vacuum type with the green grid wire and the red plate wire, and the candelabra-base connection for the filament.

But in the course of my detector experiments, I should not omit reference to one experiment in the field of the electrolytic—and finance. Wollaston wire in San Francisco was only obtainable at a dealer's on lower Market Street—25c an inch. Shipments were received irregularly from the East. So one day, some of the local hams pulled a "Black Friday" on the rest of the fraternity and cornered the local Wollaston market. It probably cost us all of $3.75 but the pool sold out at 100% profit.

One of our number was really fired with genius. When the de Forest audions first made their appearance on the Coast, along about 1913, he noticed that in the cold morning hours, his detector was never as sensitive as it was when he had gotten his "shack" heated. He discovered that his audion was a little on the hard side and the warming of the surrounding air occluded enough gas molecules from the inner side of the glass envelope, or perhaps the metal electrodes, to lower the vacuum to the optimum point for maximum sensitivity. Thereafter, a local glass blower pumped each of our audions just past the critical stage, we operated them suspended downward in small beakers of oil, and with a Bunsen burner brought them to—and maintained them at—a degree of sensitivity which made our DX records the envy of the less erudite hams of our acquaintance.

Next followed the use of a U-shaped magnet around the outside of the audion for so directing the ionic (or electronic?) stream within the electrode field as to secure still greater sensitivity. I think I contributed a monograph in this field of "research" to *Modern Electrics* in 1914. I know that when the War came, and I packed up to join the Navy, the old magnet went with me and saw active duty at NPL to the astonishment of some hard-boiled sailors who wouldn't believe what it could do for a weak signal until I showed them.

Those were the days before amplifiers when we had to extract the last ounce out of the incoming signal—when poor or careless engineering couldn't be compensated-for by just adding another tube. Litz wire in receivers and high-impedance Brandes "cans" helped, but many were the schemes we tried, to get out that "last drop".

Now we fight for straight-line performance, ±0.5-db. from 30 to 16,000 cycles. The thrill of music of the great masters has replaced the thrill of the faint whisper of a 240-cycle note in the 'phones. Today, we fight acoustics and control line cut-off and long-line phase distortion, when formerly we fought static and interference from a "rock-crusher" a few miles away. It's still the same old game but perhaps just a little less selfish, for in the old days only the ham of the family got the thrill; today, the entire family can respond to Wagner or to Beethoven—to both under the incomparable Toscanini.

ROBERT H. MARRIOTT—

Consulting Engineer

FORTY years ago, Robert H. Marriott became interested in "wireless", studied it, and experimented with Hertz equipment. In order to send and receive, he developed automatic recording apparatus. He recalls the first commercial receiver, early news transmissions and police alarms. Mr. Marriott was the first President of the I.R.E.; is now a consulting engineer.

AMONG the many who are entitled to credit for radio, Heinrich Hertz seemingly has not received enough. Hertz was the first to take up the Maxwell theory and from that theory to set about designing, building and operating devices that would demonstrate radio transmission, reception, waves, wavelengths, frequencies, tuning, reflection and refraction.

Following the publicity about Marconi in the Spring of 1897, I asked Ohio State University how I could study wireless. They advised me to take the General Science Course—majoring in Physics and specializing in wireless—which is what I did from 1897 to 1901. The Hertz radio equipment was what I used first and I understood it had been there for several years. I presume that there were copies and modified copies of the Hertz radio equipment in many colleges, in increasing numbers, after the Hertz 1887 disclosures.

Hertz transmitters and receivers were made small enough to be used in either the horizontal or vertical position with reflectors and lenses on lecture tables before classes, with the transmitter and receiver on the same table. *The radio frequencies with Hertz apparatus, between 1887 and 1897, probably averaged around 150 megacycles.*

From 1897 on, radio expanded beyond the walls of college physics laboratories. The antennas became higher, the waves became longer, the powers became greater, the detectors became more sensitive and the communication distances became longer. Maxwell-Hertz radio that had been academic for several years was trying to become commercial with Marconi and others. However the publicity about Marconi did not cause a general stampede to the radio field. No fellow student would take up radio with me, therefore, when I passed from operating the apparatus where both the key and detector were within reach to longer ranges, *I had to arrange the transmitter to be keyed by a clock and include a tape recorder in the portable receiver!*

The X-ray also began to be a commercial product in about 1897. Induction coils (later to be popularly referred-to as "spark-coils") that were made for X-ray were also used for radio. The first factory-made radio receiver for experimental and private use, that I saw, was purchased during the academic year 1899-1900 by Ohio State University from the J. H. Bunnell Company and was made by the United States Electrical Supply Company, both of New York.

Disbelief in radio was common and persistent. In 1902 newspapers published test messages which they had sent over the Catalina Island to the California mainland radio circuit, a famous physician was obtained quickly by radio for an Avalon guest and the result of the Jeffries-Fitz-

Reminiscences

simmons fight was reported by radio to Avalon many hours before that news could have been delivered otherwise. Those successive demonstrations only made the disbelievers talk more and louder.

But one morning 2 negroes left Avalon on the 5 o'clock boat before it was discovered that they had taken the cash and expensive wine from the Metropole hotel bar. Avalon citizens believed from previous experiences that such thieves would not be caught because the infrequent boats provided such a head start. The hotel manager sent a radio, with the result that the police met the colored men when the boat docked at San Pedro. That radio demonstration silenced disbelievers at Avalon and they used that radio circuit for 21 years, until cables were substituted.

McMURDO SILVER—

Pres., McMurdo Silver Corp.

McMURDO SILVER, guiding spirit of the corporation which bears his name, counts 26 years of activity in the radio field, dating from the time when apparatus included such accessories as gas burners and permanent magnets, and such parts as yard-long lengths of cardboard tubing. Mr. Silver is one of the pioneer manufacturers of high-quality receiving equipment.

IN this, the 50-year JUBILEE NUMBER of *Radio-Craft*, the writer is somewhat amazed to realize that he can actually look back over an active 26 out of this total of 50 years.

Very well does he recall his introduction to wireless in 1912; his first and fruitless attempts to build a spark coil, and his final reasonably successful attempts to build a single-slide tuning coil and crystal detector.

"Much water has flowed under the bridge" and it is almost impossible to project oneself back a quarter of a century and then try and look forward at radio. It seems equally difficult to attempt today to look forward another quarter- or half-century.

The writer distinctly recalls his first thrill at finally hearing POZ with a "1-lung" (only one filament still burning) audion bulb, flanked by a Bunsen burner and an old telephone magnet to increase its sensitivity; and the difficulty had in keeping untangled from what seemed endless lengths of cardboard carpet-tubing, for the winding of which the local electric shop had given its last inch of No. 36 S.C.C. copper wire.

Equally thrilling was the occasion in the fall of 1914 when Hobart College played against Cornell's football team at Ithaca, actually won and no one in Geneva, New York, knew this because a storm had cut off telephone and telegraph communication. For the first time the villagers appreciated amateur wireless activities, for despite the elements, news of the transcendent victory was brought in over a Colby tuner and the same 1-lung audion bulb.

The memory of almost any engineer will yield innumerable anecdotes and one man's reminiscences would almost fill this whole magazine, so the writer will sign off with only one more.

In either 1920 or 1921, the Aurora Borealis was visible early one evening in New York City. On that evening, he had gone to help a brother amateur get his VT-2 tube to oscillating in one of the new fangled tube transmitters, on Broad Street, Newark. Arriving before dinner was over, the writer fired up the rig alone and was delighted indeed to see the old G.R. "hot-wire" ammeter climbing slowly from the hoped for 1/10-ampere to all of 1¼ amperes. Looking over the rig and out of the window, he saw the Northern Lights in the sky over the Hudson. Promptly shutting off the power, he was amazed to notice the ammeter slowly climbing to an ampere-and-a-half, and even more slowly drop to zero. Needless to say, experiments were discontinued that evening.

What will come in the next 50 years—maybe television?

ISIDOR GOLDBERG—

Pres., Pilot Radio Corp.

PILOT parts were known to all experimenters and constructors in the latter part of the 1920's. This astute business man had the foresight, in those days, to manufacture the greatest variety of radio equipment at prices designed to have popular appeal. Today, he heads the internationally famous Pilot Radio Corporation. The letter below is addressed to Hugo Gernsback.

WHEN I received your letter of recent date informing me of your JUBILEE SOUVENIR NUMBER, it carried me back to the days when you and I became associated in 1908 in the manufacture of wireless apparatus.

Those were the days of the sliding tube condenser, ball-bearing tuning coil, coherer detectors, electrolytic detectors and the inevitable cat's-whiskers. Well do I remember the 1-, 2-, 3- and 6-inch spark coils, each weighing it seemed to me about a ton apiece, being sold as rapidly to the experimenters of that time as tubes are now being sold to Service Men. Every inch of spark coil power meant 5 miles of transmission and how desperately anxious all the boys were to accumulate enough to get larger and larger spark coils. Now you know why Pilot Radio claims 29 years' experience in the Radio Art.

I want to go on record as saying that I don't think credit enough has been given you by the radio industry for the pioneering work you have done in the Art. You are responsible for the vast number of experimenters in the wireless field and without them, we wouldn't have the radio developments which we have today.

May you continue your good work for many years to come; I wish you continued success.

ARTHUR A. HEBERT, *W1ES*—

Treasurer, The American Radio Relay League

THE treasurer of the American Radio Relay League, Arthur Hebert, has been an amateur since 1902. His first real transmitter was bought at the old E. I. Co., and operated under the call AH before Federal licensing provisions went into effect. Mr. Hebert has operated 2ZH and 2MP in addition to his present W1ES. As an amateur, he is a firm believer in "ham" communications.

IT is just 35 years ago that I, as a young railroad telegrapher, in Providence, Rhode Island, became infected with the "wireless bug" while trying to make a coherer

Of Old-Timers

"decohere" and enable one to receive wireless signals from the Massie Wireless station in that city,—and the bug has never left me.

Shortly thereafter, I accepted a position in New York City, and there found kindred spirits, but the business of making a living put a damper on much activity, but enabled one, nonetheless, to browse about steamship piers, "DF" at Manhattan Beach, "WA" at Waldorf-Astoria Hotel (the famous Col. Wilson's stock station) and other commercial stations, becoming acquainted with the "wireless operators", which kept the interest alive until 1908 when a momentous decision was made of acquiring a real transmitter.

What a thrill while walking through Fulton St. during the noon hour in picking up a little magazine named *Modern Electrics* to discover after reading a few pages that there actually was a store in New York City where "wireless" parts could be bought. The desire to see the place that same noon could not be resisted and on we walked through Fulton, Greenwich and West Broadway, and there on the 2nd floor of an old building we met for the first time, the Father of Amateur Radio, Hugo Gernsback, the publisher of *Modern Electrics* and the owner, if memory fails not, of the famous E. I. Co.; and at the same time it was my pleasure of meeting Victor Laughter, author of a book on wireless. The amateur spirit was there as we have it today and what a "talkfest" we had,—it resulted in my parting with a $20 bill and leaving the place 2 hours later with a 1-inch spark coil, a 1-slide "tuner" guaranteed to receive 1,000 miles—it had 1,000 turns of black enameled wire on it—, and a pair of headphones which we still have.

During the next 3 years we find ourself "puttering" and feeling very much discouraged because the "Molybdenum" detector would not bring in signals; the "Electrolytic" detector "ate up" the "Wollaston" wire at the rate of almost 1 inch per hour and as this wire cost us $1 per inch, it became too much of a luxury. But when we discovered a source of supply where we could buy Galena at 25 cents per pound, there was a silver lining in our radio world.

The detectors made with galena were so good and sensitive we found ourselves in the position of being able to copy the Navy Yards all along the Atlantic Coast and, many times, assist in copying messages which Navy and commercial operators had difficulty in receiving. In that manner our circle of friends amongst the wireless operators grew, but once in a while irritation would develop, particularly when our 1-kw. spark transmitter caused interference and we were asked to "stand-by", but it had to be done politely; otherwise the poor Navy, ship or land station operators were in hard luck as our signals were stronger than theirs.

When the laws of 1912 were passed by Congress, and the administration placed with the Bureau of Navigation, Dept. of Commerce, station and operator licenses became obligatory, the old "AH" call changed to 2ZH; after the war to 2MP and as such was as well known as W1ES has been since 1923 when we moved to the first call area.

With the advent of government supervision rapid changes took place and Amateur Radio—through the efforts of the American Radio Relay League, founded in Hartford in early 1914, by the famous inventor and amateur, the late Hiram Percy Maxim—made astounding progress in the field of communication and experiment.

Whom do we find prominently connected with the radio industry today but the men who were radio amateurs 20 or more years ago? Amateur radio is still the great training school of the engineer of today—and tomorrow.

The story of amateur radio is one of service to the government and to communities in time of disasters.

The early pioneers deserve great credit for developing a "hobby" which is the admiration of the world, and where it may lead to, with television just around the corner, I shall leave to the prophecy of Mr. Hugo Gernsback, whose many editorials have been so far ahead of the times.

WILLIAM H. PRIESS—

Pres., International Television Radio Corp.

PRIESS' work on the development of reflex receivers, while with Kolster and when manufacturing sets under his own name, is well known to many old-timers. His more recent work in television, described in recent issues of Radio-Craft *magazine, is becoming outstanding and popular. His reminiscence has exceptional value. (Photo is from a painting by M. Kassel.)*

THE summer of 1911 found me working in the electrical and blackface gang at the old Waldorf-Astoria. A job requiring a strong back.

Tesla, the Montenegrin genius of Mephistophelian get-up lived on the 11th floor. This man with his induction motor created the alternating current industry, and dreamed of the Earth as you and I would of an apple, which he would activate from a central point to drive ships and trains, give light and heat everywhere without wires. He played with millions of volts and made property cheap in the vicinity of his laboratory. He gave us the multiple spark-gap, and Tesla transformer. His roost was generally avoided by us all, and we tip-toed by it as a youngster by a graveyard at night. On the roof of the Waldorf, a wireless station crackled messages to the ships at sea.

Marriott, of the old United Wireless, the grand-daddy of the Federal Communications Commission field service. I recall my pride in carrying his wavemeter, Braun hot-wire ammeter and voltmeter, in 1912 on ship inspections. Can you remember the racket and the roar of the early commercial wireless sets doing their job at an efficiency of one-half of one per cent? Do you remember the Fleming Cymometer?

Bucher, teaching the Marconi School, in a dismal loft at this time. We two aimed at writing a correspondence school course on wireless, with directions for using the keen stuff Electro Importing Co. was offering. The talented Bucher became the first sales manager for the RCA and later its Vice-President.

Shoemaker, with the largest cud of tobacco ever carried in the mouth of a white man, performing tricks with link circuits. The old-timers remember how he planned a reserve flag wig-wag system to make sure there would be no slip in the advertised first wireless reporting of the America's Cup Yacht Races.

Austin, his mastery of meter and buzzer, and his range data established by the cruise of the *Salem* and the *Birmingham.*

Pickard, the cultured, the foremost scholar of the business and a veritable walking engineering encyclopedia, also an aspirant for wireless cup-race honors. He is of the crystal detector and I.P.76 receiver fame. Have him show you the repellent picture, taken 30 years ago, he has of himself, lolling in a bunk on his reporting tug, bearded and as nondescript in appearance as a tramp on a hand-out line.

Sarnoff, pounding brass in a two-by-four station on top of Wanamakers. Now President of RCA.

Goldsmith, teaching radio and banging out the *Proceedings*

(Continued on page 620)

THE STORY OF

In times of stress, when all other means of communications fail through the action of fire, flood or storm, the radio amateurs come through! Everyone knows of their sensational achievements in recent years. Yet hardly a person

RCA

Old-time amateur station of Mr. Willenberg-1909-'10, using spark transmitter & Leyden jars.

RCA

R.C. of A. amateur station at Hotel Ansonia, transmitted to Admiral Fletcher's flagship in 1909

IT was in 1908," says Clinton B. De Soto, in a book published by the A.R.R.L., "that Hugo Gernsback, a clever Belgian* promoter, manufacturer and shopkeeper on Fulton Street in New York City . . . . began the publication of the magazine *Modern Electrics* . . . . The contents . . . . soon became predominantly radio in character."

A little farther on, Mr. De Soto continues, "The Wireless Association of America was a child of Hugo Gernsback, publisher of *Modern Electrics*. In early 1910 the first *Wireless Blue Book* of the Association appeared, dated 1909. It listed 90 U. S. amateur stations . . . . together with the call letters used, wavelength in meters and spark-length of the induction coil . . . ."

To Mr. Gernsback, thus, may be attributed the development and organization of amateur radio, in a great degree. Without his foresight and friendly interest, it is highly doubtful that the "hams," as the amateurs call themselves, would ever have reached their present high degree of development. His early publication of a magazine devoted to wireless and scientific amateurs stimulated and encouraged the youths to interest themselves in fascinating and intellectual pastimes; later his fostering of the "amateur association" movement resulted in an organization destined to play no small part in the growth of amateur radio.

Amateur radio may be said to have received its first impetus when the late Guglielmo Marconi, then a young Italian experimenter, received 3 dots sent across the Atlantic Ocean on December 12, 1901. The apparatus he used was crude, and amateur scientists, inspired by his achievement, built similar set-ups and proceeded to try their hands at sending and receiving.

There were no licenses required in those early days. The sole requirement for amateur radio activities was the equipment. If one could afford to buy the parts and had the skill to use them, he could go on the air, using any frequency or power his heart desired. Sometimes his heart desired a frequency adjacent to one used by the Navy, or commercial wireless companies, after these had come into existence. If so, it was just too bad, for the amateur was under no obligation to shut down his service or shift his wavelength if he was interfering with other communications; it was a matter left to his own discretion. This condition held from the beginnings of radio until August 17, 1912, when President Taft signed an act to regulate radio.

Although 28 bills to regulate radio had been introduced in Congress during the preceding 10 years, this was the first to affect the amateurs—and affect them it did! It made licensing mandatory

*Mr. Gernsback is a native of the Grand Duchy of Luxembourg, not Belgium. He is now editor and publisher of *Radio-Craft* and other magazines.

RCA

Miss A. McConaughy, 13 years old (at that time) operating her licensed amateur wireless station in 1910.

RCA

Princeton student amateurs playing chess by wireless about 1912

OTM.

Portable amateur radiophone station in Indiana (owned and operated by Continental Elec. Co.) about 1915.

knows of the amateurs' less spectacular but equally important contributions to radio. The annual Paley Amateur Award (shown at right) is presented to the amateur who "has contributed most usefully to the American people".

for all operators, including the amateurs, whom it banished to the limbo of the lower wavelengths, *then considered useless* for long-distance transmission. It regulated their transmission to 200 meters or less, their transformer input to a maximum of 1 kilowatt.

Amateurs immediately began applying for licenses, and by the end of the year approximately a thousand had secured theirs. Another year saw another thousand licensed amateurs, and by the middle of 1914, more than 5,000 were licensed. It is estimated that an equal number of stations were operating without licenses, on the old theory that "to err is human; to get caught is crime." Even the licensed hams did not permit their activities to be circumscribed by the law. If they felt like operating on 300 or 350 meters, they did so, and if the signals did not get out well enough with 1 kilowatt input, they doubled or trebled the power.

Still the amateurs craved distance communications, and in 1912 Edwin H. Armstrong, 22-year-old ham, played around with a de Forest audion and made regeneration work. De Forest had discovered regeneration a year before, but had not as yet put in a patent claim for it. To this principle, many trace the use of oscillating tube circuits for both transmission and reception.

In March the following year, the amateurs established the tradition that hams have carried on ever since—that they can be depended upon to "come through" in times of emergency. Amateur stations throughout the country, and especially those at the University of Michigan and Ohio State University, carried on communications for a large section of the mid-West when all other forms of communication were rendered inoperative by severe storms.

During all these years there had been many radio clubs throughout the country, and in 1914 Hiram Percy Maxim, president of a Hartford (Conn.) radio club, wanted to send a message to another amateur who was beyond the range of his transmitter. So he called another amateur who was nearer and had him relay the message. Though message relays had often been conducted before, this was the first time that they had given anyone a really new idea. The idea was for the establishment of the American Radio Relay League, the first steps toward which were taken on April 6, 1914. By September it had 237 member stations in 32 states and Canada, paying no dues save in enthusiasm. In early 1915, the League—the A.R.R.L.—separated from the Hartford club and was legally incorporated as a separate organization. The first official A.R.R.L. magazine was published in December of that year.

The war had broken out in Europe, in the meantime, and the

(Continued on page 618)

P.

Mobile Amateur Radiophone Station, owned and operated by Mr. L. G. Pacent in 1919.

RCA

Armstrong and other amateurs sent wireless signals from [illegible] (trans-Atlantic amateur tests) in Dec. 1921

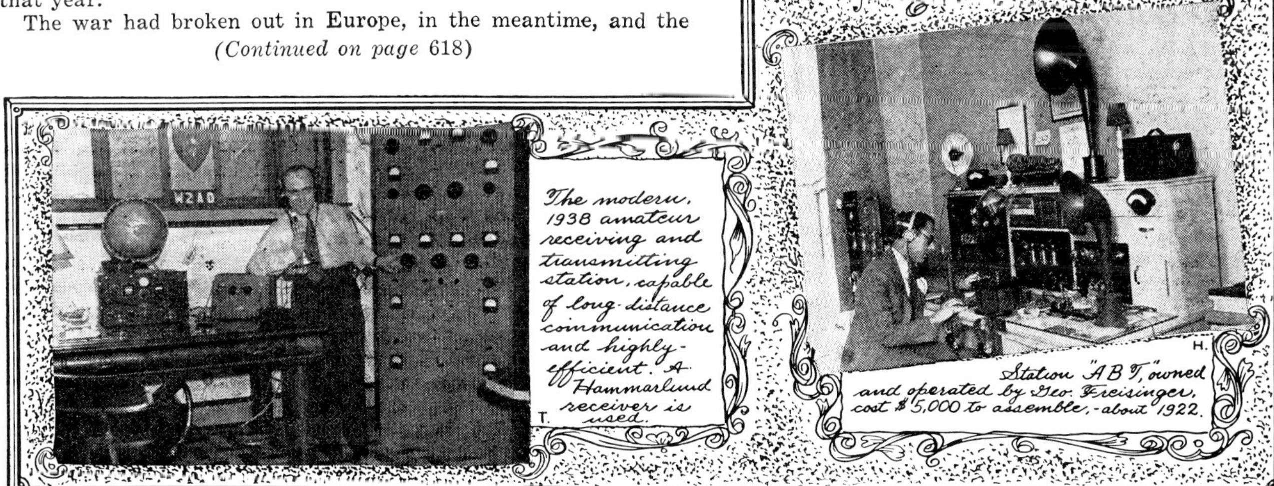

T.

The modern, 1938 amateur receiving and transmitting station, capable of long-distance communication and highly-efficient. A Hammarlund receiver is used.

H.

Station "ABT," owned and operated by Geo. Freisinger, cost $5,000 to assemble,-about 1922.

RADIO PARTS

Electrolytic Detectors about 1902-'03 (N.E.S.Co) G.C.

Coherer Detector used from 1896-1902 M.W.

Electrolytic Detector used from 1905-1924 (N.E.S.Co) G.C.

Crystal Detectors used from 1907-1924 P.

Thordarson "Flexible" Transformer about 1908 T.E.

Heterodyne Headphone for Fessenden circuit, about 1909 G.C.

Selenium Cell sold by E & Co about 1910 P.

Old Transmitting Parts; Helices and changeover switches, from 1910-1920. P.

Early Leyden jars and Mica condenser, about 1912 & 1918 C.D.

Multiple Crystal Detector Stand, about 1913. G.C.

IF you want to go back for a survey of the earliest radio parts, you will have to refer to the work of a Washington, D.C., dentist, Dr. Mahlon Loomis, who used such items as kites, copper-screen antennas and galvanometers to send signals a distance of 18 miles—in 1869! Later he turned his thoughts to a buzzer and a spark-gap, which were to be actuated by the difference in potential between the earth and the higher atmosphere, where the kite held the antenna aloft. There is no record that he ever performed this latter experiment.

The open rings used by Heinrich Hertz in 1887-'88—a low-value inductance in series with an air gap—to demonstrate transference of energy from a circuit in which a spark was produced, might be termed another form of early radio part. But that likewise antedates broadcasting, as we know it, by so much that it may be considered the Stone Age of Radio.

EARLY TYPES OF RADIO COMPONENTS

More recent are the *electrolytic detector* and the *coherer*. Although the latter has been traced back as far as 1835, it was not usable until the *decoherer* (a tapper similar to an electric bell) was invented to jar the filings, in the tube, apart after they had been caused to cohere (stick together) by the incoming signal. Marconi patented the coherer, which Sir Oliver Lodge had suggested as a possible radio device a few years earlier, in 1894. The decoherer in its usable form was introduced in '96.

These devices co-existed for some time with the electrolytic detector, until 1907, although the electrolytic detector was the more popular of the two because of its greater stability and efficiency. The Fleming 2-element rectifier, invented in 1904, and which we call the *Fleming valve*, while a revolutionary improvement and destined to be the basis for a still more remarkable achievement in radio, was never popularly used for the reason that the "diode" was too insensitive to weak impulses.

Pickard's invention of the *crystal detector*, in 1907, was a great step forward in this direction, since it provided a very simple and economical detector which was the most efficient of all until the "triode" was made commercially available. The crystal detector was popular from the year it was discovered until 1924, for it permitted the construction of simple, practical and economical receivers. Many various metal ores were employed in this type of detector, some of which were *carborundum, galena, silicon, iron pyrites, copper pyrites* and *zincite*. The coarser the grain of the crystal, the better it was supposed to act as a detector. Some manufacturers went so far as to produce a detector with a multiple number of crystal cups, so that any one might be rotated into position and the cat-whisker placed in contact thereto in quick order.

About that time, primitive inductances were made. These were simply long coils of wire, with two sets of taps. The first 10 turns were each tapped; every 10th turn from there on was also tapped. In this way, the experimenter could tune by cutting any number of turns into the circuit. These were still in use in the late 'teens and early 'twenties, when many listeners were still

Of Yesteryear

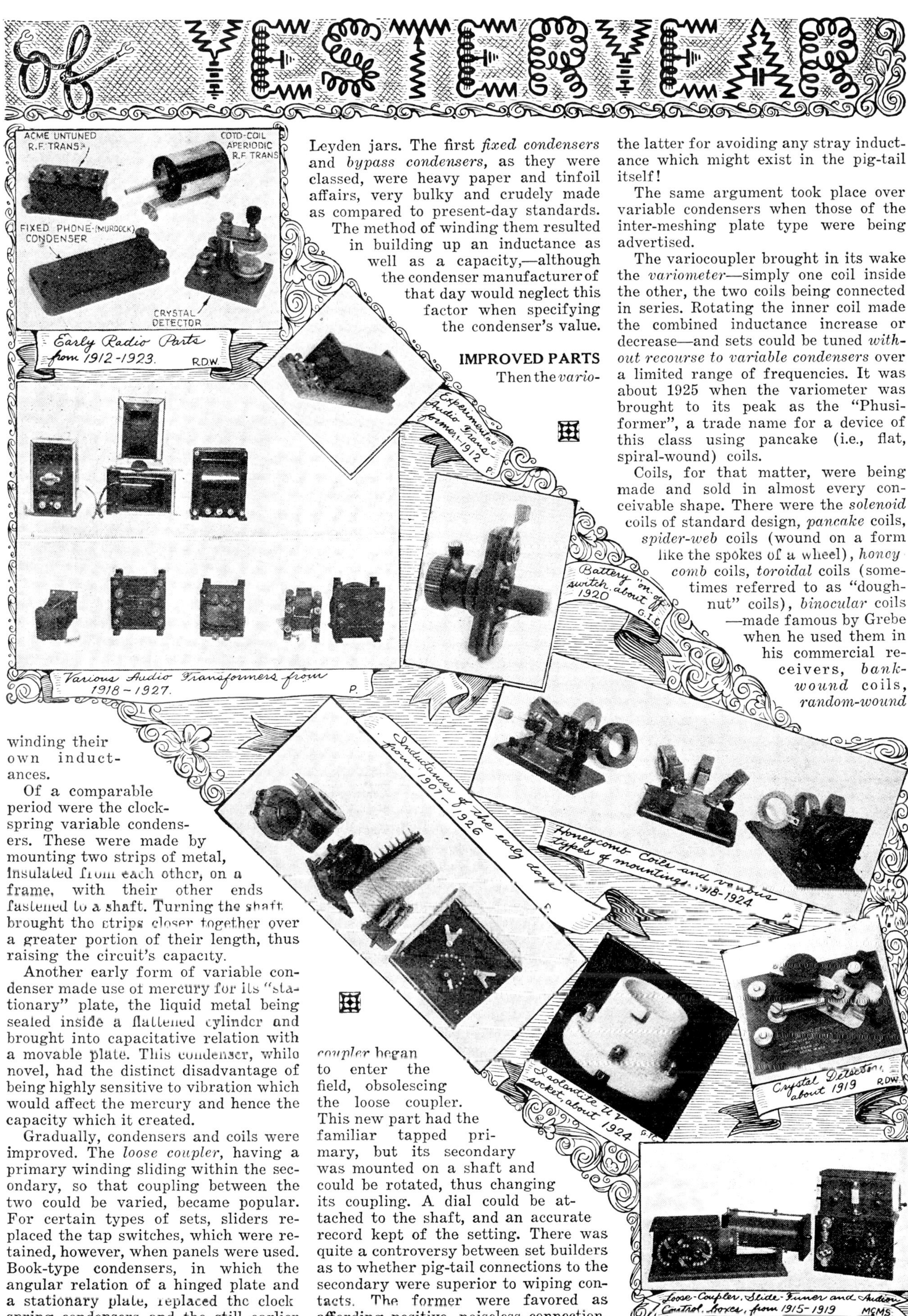

winding their own inductances.

Of a comparable period were the clock-spring variable condensers. These were made by mounting two strips of metal, insulated from each other, on a frame, with their other ends fastened to a shaft. Turning the shaft brought the strips closer together over a greater portion of their length, thus raising the circuit's capacity.

Another early form of variable condenser made use of mercury for its "stationary" plate, the liquid metal being sealed inside a flattened cylinder and brought into capacitative relation with a movable plate. This condenser, while novel, had the distinct disadvantage of being highly sensitive to vibration which would affect the mercury and hence the capacity which it created.

Gradually, condensers and coils were improved. The *loose coupler*, having a primary winding sliding within the secondary, so that coupling between the two could be varied, became popular. For certain types of sets, sliders replaced the tap switches, which were retained, however, when panels were used. Book-type condensers, in which the angular relation of a hinged plate and a stationary plate, replaced the clock spring condensers and the still earlier Leyden jars. The first *fixed condensers* and *bypass condensers*, as they were classed, were heavy paper and tinfoil affairs, very bulky and crudely made as compared to present-day standards. The method of winding them resulted in building up an inductance as well as a capacity,—although the condenser manufacturer of that day would neglect this factor when specifying the condenser's value.

IMPROVED PARTS

Then the *vario-coupler* began to enter the field, obsolescing the loose coupler. This new part had the familiar tapped primary, but its secondary was mounted on a shaft and could be rotated, thus changing its coupling. A dial could be attached to the shaft, and an accurate record kept of the setting. There was quite a controversy between set builders as to whether pig-tail connections to the secondary were superior to wiping contacts. The former were favored as affording positive, noiseless connection, the latter for avoiding any stray inductance which might exist in the pig-tail itself!

The same argument took place over variable condensers when those of the inter-meshing plate type were being advertised.

The variocoupler brought in its wake the *variometer*—simply one coil inside the other, the two coils being connected in series. Rotating the inner coil made the combined inductance increase or decrease—and sets could be tuned *without recourse to variable condensers* over a limited range of frequencies. It was about 1925 when the variometer was brought to its peak as the "Phusiformer", a trade name for a device of this class using pancake (i.e., flat, spiral-wound) coils.

Coils, for that matter, were being made and sold in almost every conceivable shape. There were the *solenoid* coils of standard design, *pancake* coils, *spider-web* coils (wound on a form like the spokes of a wheel), *honeycomb* coils, *toroidal* coils (sometimes referred to as "doughnut" coils), *binocular* coils —made famous by Grebe when he used them in his commercial receivers, *bank-wound* coils, *random-wound*

RADIO PARTS

Types of Variometers,-1917 to 1925 R.D.W.

A 10-to-1 ratio dial-about 1924 R.D.W.

Compression-type Variable Condensers,1920-'24 R.D.W.

Early Filament Rheostat,-1922-'25 G.I.

Slow-speed single tuning condenser,-1926 G.I.

Early Audio Transformer,-1923 P.

Old-time Army-Navy Changeover Switch -1916-'20 A.R.C.

coils and a year or two later, *lemniscate* coils, wound in the form of the figure 8.

THE "LOW-LOSS" ERA

Early in 1925, the industry became "loss conscious", electrically speaking. Constructors and engineers began to worry about the R.F. resistance in coils and condensers, and from then on everything had to be *low-loss* to be acceptable. Dielectric end-plates which had supported the bearings in variable condensers were discarded, the insulation being moved to the side, where it supported the stator plates. The substitution of metal end-plates was of even further advantage, for it enabled the set builder to reduce the effects of body capacity to a minimum while tuning. One manufacturer of condensers held out to the end, advertising, "Don't be fooled by 'Low Loss' BUNK!", but the public would not heed, and condensers of the "low-loss" type are in use to this very day.

Coils, too, came in for their share of attention from the efficiency experts. The spider-web coils, stuck into position with cement, had a spell of popularity. The solenoid coils were liberally coated with similar "dope" and slipped from their winding forms, of which only vestiges were permitted to remain as terminal and mounting supports.

One manufacturer even went so far as to silver plate the primary of his 3-circuit tuner (in this case, a variocoupler with additional fixed winding to act as antenna coil) on the assumption that the greater conductivity of this metal would improve reception. Some users reported that it did!

Shortly after the "low-loss" era, began the period of condenser plate curvature improvement. Variable condensers of the inter-meshing type as first designed were of the "*straight-line capacity*" variety. With the ever increasing number of broadcast stations, it was found that on the lower wavelengths many stations were crowded into a narrow range of the tuning dial and spread too far out on the higher wavelengths.

To correct this the *straight-line wavelength* condensers were brought out, this curve being obtained by elongating the shape of the plates a bit. While the new condenser somewhat corrected the condition of crowded tuning on the low-wavelength end of the dial, it wasn't completely satisfactory for the reason that broadcast stations were being assigned wavelengths on a 10 kc.-separation frequency basis.

Hence, to get equal frequency changes over the entire tuning range of the dial, the tuning condenser had to have a straight-line frequency curve,—which resulted in an era of almost elliptically-shaped condensers, known as *straight-line frequency* condensers. One of the first and most ingenious of these was the "square law" condenser designed by Clyde J. Fitch; its triangular plates were meshed, point first, by means of an ingenious screwthread gear system and 360° dial.

However, some manufacturers saw an opportunity to cash-

Army-Navy Microphone and Headphones-1916-'20 A.R.C.

Moulded-end-plate single tuning variable condenser-1916-'24 G.I.C.

C. J. FITCH TROPAFORMER (TRIMMER CONDENSER)
R. E. LACAULT ULTRAFORMER

Early commercial I. F. Transformers-1923-'26 R.D.W.

in on the fact that many thousands of variable condensers of the older style were already in use, or else available at sacrifice prices, and figured that some device was necessary to give straight-line frequency tuning with old-type variable condensers,—and so was born the *straight-line frequency dial.* It consisted of nothing more than an eccentric-shape disc driven by a smaller disc by friction, so that slow vernier tuning was accomplished on the lower wavelengths and practically normal direct-drive tuning on the higher wavelengths.

MISCELLANEOUS PARTS

Variable resistors, too, found a ready market. Each tube, of course, was supposed to have its own rheostat for most efficient operation. Then vernier rheostats, some with extremely ingenious design for varying the resistance a fraction of an ohm at a time, were sold. These were supplanted, in the higher-price ranges, by compression-type rheostats, which used piles of carbon discs or conductive powder, the resistance being varied as a knob was screwed tighter or looser. Subsequently it was learned that filament voltage was not so critical, especially on the audio amplifier tubes, and fixed, self-adjusting filament resistors, such as the Amperite, came into general use as an effective means of compensating the change in battery voltage between charge and discharge.

Potentiometers were sold separately or combined in one instrument with a rheostat—and even grid leaks were made variable. One of the latter, sold at a high price, was actually hand-calibrated in megohms and fractions thereof, but was highly hygroscopic, and its actual readings varied with changes in weather.

All during the set-building boom, which ran from about 1922 to 1931, kits were available. Some of the earliest were reflex kits. Later the neutrodyne was sold in kit form, the basic essentials being matched *neutroformers* (coils) and *neutrodons* (neutralizing condensers).

The condenser manufacturers were quick to take advantage of this, and at least one was soon advertising sets of 2 and 3 individual, *matched* tuning condensers, the use of which was said to make the dials "log" perfectly.

At about the same time, *ganged condensers* swung into favor. Some of these used a common shaft, as at present, but others used gearing and belts to tie together several separate units. One popular model had plates of approximately square shape, which were pivoted at the corners, both sets of plates moving. In the early gang condensers, variable trimmers were nearly always added to afford finer tuning.

Vernier dials were another part that was popular to achieve the same end. Geared drives, and friction drives of various sorts were employed to secure positive action and freedom from backlash. (One prominent make of dials even had a sort of gearshift lever for varying the ratio.) Re-

(Continued on page 642)

Early condenser-ganging by means of gears—1925. G.I.C.

SPIDER-WEAVE
TOROID OR "BALLOON"
MODERN OSCILLATOR COIL

Evolution of tuning coils—1924-'38 R.D.W.

Early D.C. type "Trickle" Charger 1922-'28 W.L.

(LAMP SOCKET)

Electrolytic-Type Trickle Charger—1926 X.L.R.

"Low-loss" ganged variable condensers—1928 G.T.C.

Neutrodyne Coil Kit for constructors—1926 W.M.C.

Early forms of Variable Condensers—1912-'22 P.

Novel Two-gang variable condenser—1926

"B" Eliminator for replacing "B" batteries—1926 J.T.R.

Development of

ALEXANDER GRAHAM BELL'S FIRST MICROPHONE·1875

Practical transmission of sound began with the inception of the telephone, hence it was within this industry that the first Microphone was developed. Credit for much of the improvement of this device, however, belongs to radio, as is told in this story. Loudspeakers, Amplifiers and Phonograph

SCHOLARS have long debated whether sounds exist where there is no ear to hear them. Whatever the answer, for the purpose of this article, sound shall be considered to mean atmospheric vibration at audible frequencies; i. e., between (roughly) 30 and 14,000 cycles per second.

These are set up by vibrating bodies, such as a violin string, a trumpet reed, a drum-head or the human larynx, to mention but a few of thousands. Such sounds travel for considerable distances through the atmosphere, and even in ancient times it was found that they could be directed by means of megaphones or speaking trumpets.

But men wanted other, easier means of transmitting their voices to comparatively great distances, without relying on sheer lung power. True, they had primitive "telegraphs" such as the heliograph, beacon fires and smoke signals, but they wished to transmit the spoken word.

The first telephone was invented by an English genius in 1667, who based his work on the "bull-roarer", an instrument still used by African witchdoctors and American sideshowmen. It consists of a can, one end of which is covered by a tightly-stretched membrane, with a cord fastened in the center. This is the device used by Robert Hooke in 1667 when he transmitted voice between two distant points. Hooke linked two of these instruments together with a goodly length of string, which, when stretched taut, permitted the vibrations of a voice striking one diaphragm to be transmitted along the cord to the other. (Sound waves are transmitted much farther through denser mediums —such as cord and other solid materials —than through air which is a comparatively low-density medium.)

Nearly 200 years went by before our present devices were approached. It was, in fact, 1854 when Charles Bourseul suggested that a disc, vibrated by the voice, could be used to make and break an electrical circuit to cause similar vibrations in a second diaphragm. This idea was tried by Reiss, of Frankfort, in 1860.

Another 16 years passed and in 1876 Elisha Gray, an American, invented a system which caused variations in the strength of a steady current. In the same year, Alexander Graham Bell connected two "receivers" together, so that the diaphragm of one, actuated by the human voice, generated in the pole windings a current which was adequate to cause the other receiver to operate, reproducing the vibrations.

However, the currents generated in Bell's magnetic transmitter were too weak for use over appreciable distances, so it was supplanted by a means of modulating a constant current, thus paving the way for the *microphone.*

DEVELOPMENT OF THE MICROPHONE

Early laboratory models of the microphone were simply a light base, sup-

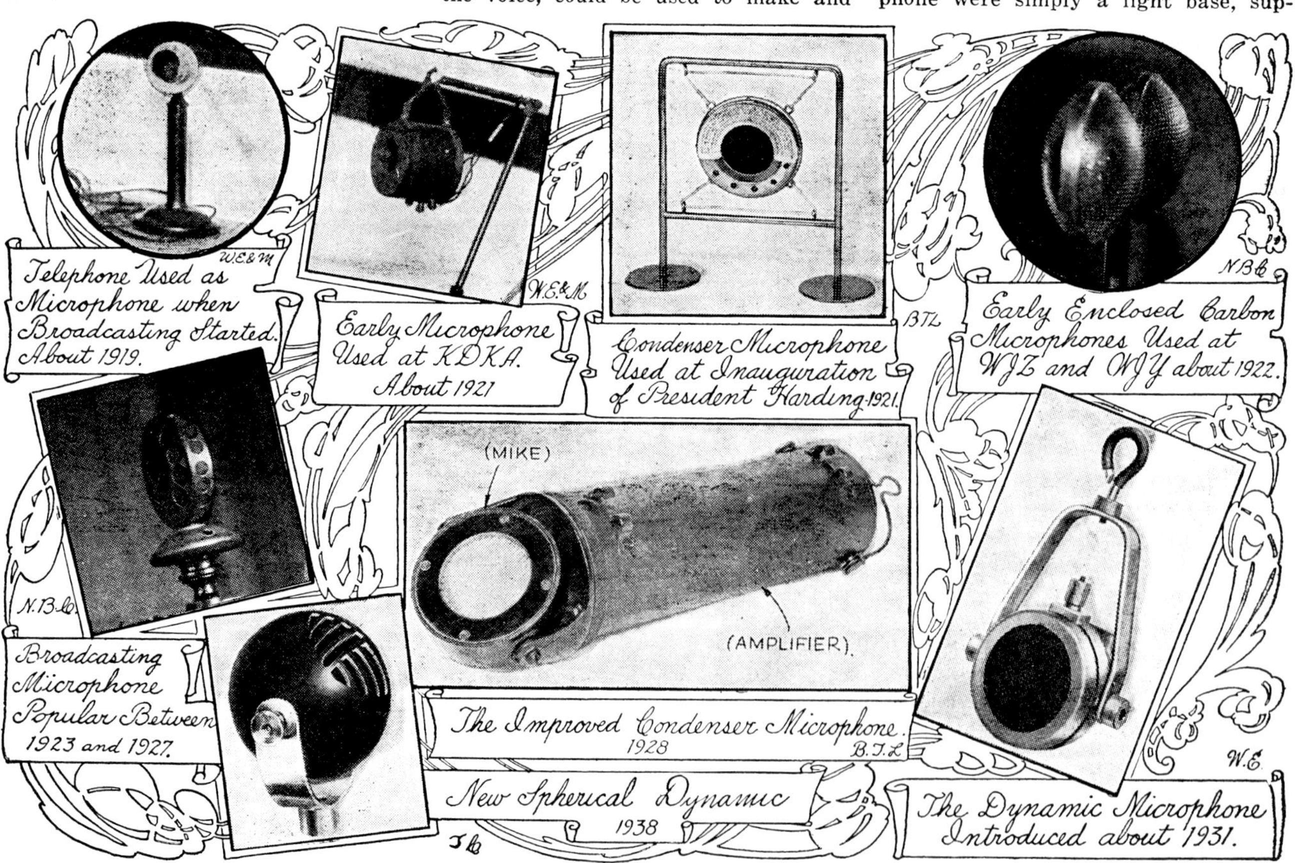

Telephone Used as Microphone when Broadcasting Started. About 1919.

Early Microphone Used at KDKA. About 1921

Condenser Microphone Used at Inauguration of President Harding 1921.

Early Enclosed Carbon Microphones Used at WJZ and WJY about 1922.

Broadcasting Microphone Popular Between 1923 and 1927.

The Improved Condenser Microphone. 1928

New Spherical Dynamic 1938

The Dynamic Microphone Introduced about 1931.

Sound Equipment

2 ELECTRO-MAGNETS CONNECTED IN SERIES

SOFT IRON DIAPHRAGM

ELISHA GRAY'S MAGNETIC SPEAKER-1879

Pickups are exclusively products of the radio laboratories. Without vacuum tubes we would have had no amplifiers,— no loudspeaker reception. The crude apparatus depicted should amaze all those who are familiar with present-day developments in high-fidelity radio and sound equipment.

porting two or more carbons in light contact with each other, and in series with an electrical circuit. Thus when the base was jarred, the contact and hence the electrical resistance between the carbons varied, causing similar variations in the current flow.

Crude and insensitive as this device was, it formed the basis of the modern carbon-type microphone, which merely affords multiple instead of one or two carbon-to-carbon contacts. The *single-button* type consisted of a fixed carbon disc, with a diaphragm-impelled carbon disc suspended a slight fraction of an inch away from it, the space between them being filled with loosely-packed carbon granules. This microphone is sensitive, but does not cover a sufficiently wide range of audio frequencies. It was largely supplanted by the *double-button* "mike", which has the diaphragm supported between two such buttons. This, while affording better tone quality, still retained many earlier defects, such as variation in its characteristics caused by changes in humidity, noisy operation (carbon noise), carbon "packing", etc.

Soon the condenser microphone, of low power output but wide frequency response, was developed. This was a 2-plate condenser, the diaphragm composing one of the plates. It was popular in the late 1920's. A preamplifier (usually built-in), connected between the regular amplifier and the condenser microphone, was soon added to boost the weak output. The ribbon-type or velocity microphone became popular about 1931, employing principles similar to the "magneto" microphone which Bell had used in his early experiments. Shortly after, the crystal mike made its appearance and at this time is the greatest favorite because of its relatively higher output when compared with other types—excepting the carbon microphone. The dynamic mike was used many years ago, then dropped, and lately has reappeared to compete with today's velocity and crystal units. Each have their special features and following, and are individually adaptable for special work.

A connecting link between microphones and loudspeakers (or reproducers) appeared in the June, 1912, issue of Hugo Gernsback's *Modern Electrics* magazine. The output of a crystal-detector wireless receiver was fed into a single phone. On the diaphragm of this phone, and connected to a separate battery, was a small cup of mercury, touching the surface of which was a carbon rod, connected to one side of a low-impedance phone unit, the other side of which completed the battery circuit. A horn was positioned on this phone unit, to act as a loudspeaker.

DEVELOPMENT OF LOUDSPEAKERS

The first reproducers, then, were simply horns positioned correctly in relation to simple magnetic telephone units. Such horns were rather costly and for this reason radio fans were inclined to improvise sounding cham-

The Early Loudspeaker. Headphones Clamped Over Tin Horn. 1921-1923 J.T.R.

Magnovox Loudspeaker. First to use Dynamic Coil Principle. 1922-'26 M.C.

Familiar Western Electric Horn Loudspeaker. About 1924. B.L.L.

Rhamstine "Needle-phone" Phono-horn Adapter - about 1925. J.T.R.

RCA 105 Powered Dynamic Speaker and Amplifier. About 1928. G.E.

3 Ft. Cone Speaker, Western Electric 1927-'8 W.E.

A Modern P.M. Dynamic Speaker - 1937-1938.

A Novel Phono-horn Adapter Unit. 1925 A.C.

Fibre Composition Horn Speaker for Better Tone S.C.

Airplane-Cloth Magnetic Speaker. Popular about 1929

Development of

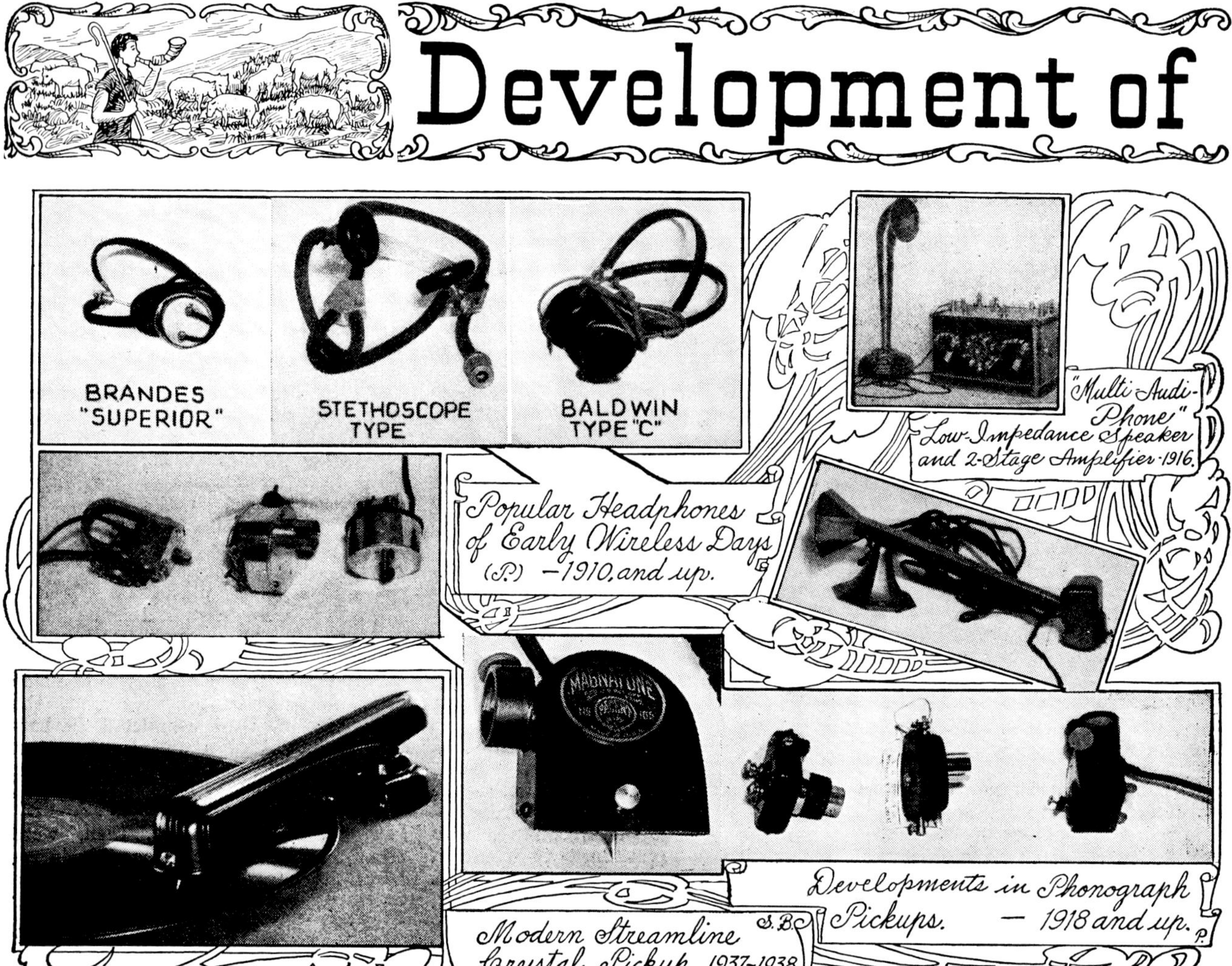

Popular Headphones of Early Wireless Days (P.) —1910, and up.

"Multi-Audi-Phone" Low-Impedance Speaker and 2-Stage Amplifier-1916.

Developments in Phonograph Pickups. — 1918 and up. P.

Modern Streamline Crystal Pickup. 1937-1938 S.B.

bers, using bowls of various sorts, to reflect and "amplify" the sound.

The main idea of a speaker was to enable a large number of persons to listen to a message or broadcast simultaneously, and simple gadgets were invented to do this without the need for speakers, as amplification was still inadequate. Among these devices were phone caps attached to a multiplicity of stethoscope tubes; plugs with terminals to permit the attachment of two or more pairs of phones; plugs incorporating switches so that phones might be connected in series or parallel; multiple binding posts (one type of which was simply a short length of coil spring), etc. However, though these served their primary purpose, they did nothing to relieve ears crushed by headphones. Therefore the search for a loudspeaker continued.

One of the earliest commercial types was merely a horn, with a place provided for attaching a headphone unit at its apex. Then came one that was "twice as good"—it had arms to hold 2 headphones. Made of aluminum, such horns sold for $10 and upward. Another manufacturer made a similar horn of papier mâché and advertised it as having "less resonance", and sold it more cheaply. Prices gradually declined until the components were on sale in the "10-cent" stores. Numerous variations were also produced, one of which was a bowl with an arm to hold a unit at its "focal point". Still another type attached a phone unit in place of a phonograph sound head to the tone arm of a phonograph.

But there were also deluxe speakers, two of which merit particular mention. The first, the Magnavox, because it employed the *dynamic* (moving coil in a magnetic field) principle upon which almost all present speakers are based. The second, the Western Electric *magnetic* (moving iron in a magnetic field) speaker, because it was generally sold in connection with a 2-stage audio amplifier, the basis of present-day public-address systems. In fact, it was the invention of multi-stage cascade amplifiers that made loudspeaker radio possible.

But there were other innovations in the early art of reproducers. Some of these flourished for a time, but only the dynamic speaker had a lasting influence.

There came the horn with phone unit built-in—the first real loudspeaker. Then followed the phone unit with movable pole-pieces, to adjust the air-gap between diaphragm and magnetic poles so that maximum volume and sensitivity could be obtained without diaphragm rattle on loud reproduction. There were also mica diaphragms with iron discs riveted to their centers, to improve sensitivity and tonal quality.

Then came the era of cone speakers, at first actuated by magnetic units. Cones of different sizes and shapes were in turn popular; for a time the belief existed that true fidelity could not be attained unless the cone was 3 feet in diameter. When this was proven fallacious, the field was explored further—resulting in balsa-wood, airplane-cloth and condenser-type speakers. All were based on the cone-type principle,— i. e., a large, light mass driven by a magnetic unit. The mechanisms of these cones were of the armature type, similar to those used in Baldwin phones.

Later, the cone, smaller in size, was combined with the dynamic driver—a combination which has retained its popularity ever since. Adequate response over most of the audio range is possible with this type of speaker when correctly-designed baffling is employed.

With the inception of high fidelity, however, there was a surge of interest toward crystal speakers, to be used in conjunction with reproducers of the ordinary type. Known as "tweeters", the crystal speakers are made with small cone areas, designed to emphasize the higher end of the audio

Sound Equipment

spectrum. Many persons still believe their use essential for optimum tone.

Today, the dynamic speaker of high-fidelity design is seldom larger than 18 inches in diameter. The electromagnetic field is eliminated in many cases, special magnetized alloy metal having been found to create a magnetic flux (or magnetomagnetic field) about as high as that obtained by electromagnetic means.

DEVELOPMENT OF AUDIO AMPLIFIER

The link between the detected signal and the loudspeaker is the *amplifier*. At first the output of the detector was fed directly into the reproducer. A little later, mechanical amplification was designed, but this was highly unsatisfactory, as judged by present standards.

Then, in the early part of 1912, de Forest with his associates Logwood and Van Ethen discovered the tremendous amplifying properties of the audion tube especially when used in 2- and 3-stage cascade (one stage after the other) amplifiers. They determined the amount of amplification (or the "amplification constant") that took place in each tube and explored the field of amplifying transformer design so as to increase this amplification. It was this work and his discoveries in this field that later interested the Bell System A.T.&T. Co., in the audion and resulted in the sale of the audion to this company.

After A.F. transformers were further improved, and radiotelephony and broadcasting started, early set builders had a problem on their hands—whether to use high-ratio audio-frequency transformers for high gain, or low-ratio ones for better tone and greater stability. Audio stages oscillated and the amplification became distorted if too much gain was sought, and a compromise had to be effected. In fact, at least one manufacturer brought out a transformer with tapped primary and secondary, so that the step-up ratio might be varied at will.

The invention of the "C" (or grid) battery circuit did much to improve amplifiers, for the use of this potential kept the grids of the audio tubes from swinging positive and causing distortion on loud signals. As mentioned in the article on old-time tubes in this issue of *Radio-Craft*, power tubes were another important factor in the amplifier field, but even before they came out, push-pull amplification was in use, to serve a not dissimilar end. For those who wished greater output than was possible with a single last-stage tube, yet who did not wish to incur the expense of push-pull transformers, paralleling of output tubes was employed.

Perhaps one of the greatest inventions during battery days was that of Latour, who patented the use of a single "B" or "C" battery on several stages. The more antique diagrams show a separate battery for the plate supply of each tube in a set. This patent was of particular value when eliminators and power packs came along, for it made it unnecessary to use a large number of eliminators for a multistage set.

Nor can this article conclude without mention of the Loftin-White direct-coupled amplifier, which was highly efficient when properly constructed. In it, the plate of one tube connected directly to the grid of the following tube, biasing being such that each grid was negative, each plate positive, in respect to its own filament.

Today, we have amplifiers that employ a minimum of tubes, have straight-line amplification from 20 to 15,000 cycles, extremely high power output and tremendous voltage gain without the instability or other inefficiencies that were predominant in old-time amplifiers.

Thus has been the progress of audio reproduction. What the next step will be, only time will tell. Most recent developments have been in cabinet and

(*Continued on page* 608)

Early Cascade Audio Amplifier. About 1919-1923. M.Y.

Old Time Magnavox Amplifier. Popular 1922. M.V.

Early Magnavox Amplifier. Popular about 1922-1924 M.V.

Famous Early Western Electric Amplifier. 1922-1924 B.T.L.

Aeriola Sr. Amplifier using WD11 Tubes 1924 G.C

Push-pull 1-Stage Amplifier using WD11 Tubes 1924 G.C

A 3-Stage Resistance Capacity Coupled Amplifier Popular 1924-1926 DC

Power Amplifier Stage and Power Supply About 1926. P

First Commercial Self-Powered Amplifier. About 1929

A DC Amplifier & Power Supply 1928 DC

Modern Amplifier. 1937-1938. W.R.S.

The Old

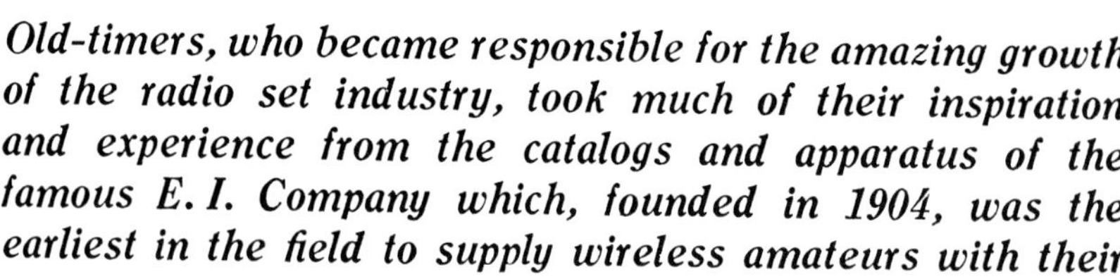

Old-timers, who became responsible for the amazing growth of the radio set industry, took much of their inspiration and experience from the catalogs and apparatus of the famous E. I. Company which, founded in 1904, was the earliest in the field to supply wireless amateurs with their

HUGO GERNSBACK

THE history of the E. I. Company really had its beginning when I was exactly 6 years old.

The entire course of my future life had its direction, probably, when my father's caretaker presented me with an electric bell, a roll of lead-encased copper wire, and an old French Léclanché wet battery. I remember quite distinctly now that the thing which captured my imagination was not so much the mysterious-looking wet battery, nor the French-made, walnut-boxed electric bell, but rather the mysterious green spark which I observed at the platinum contact points of the bell.

I also remember that this, my first introduction into the mystery of electricity, did not end there; but for many days I hooked and unhooked the battery from the bell, and later on a pushbutton was added which had been procured somewhere. From this beginning I was soon installing electric bells for friends; as, indeed, in the town where I was born—Luxembourg—there was only one electrician at the time.

(For the record, let me digress here for a moment, and state that I am neither German, French nor Belgian; though these various nationalities have been attributed to me in various newspaper and magazine articles for many years.

(The Grand Duchy of Luxembourg is a small, independent country of some 350,000 souls, sandwiched in a triangle between Germany, France and Belgium. At one time, Luxembourg was an impregnable fortress, second only to Gibraltar; in 1867 various powers [including England, France and Germany] guaranteed the neutrality of the little country if Luxembourg would raze its fortifications. This it promptly did, and this neutrality was not violated until 1914, when Germany invaded Luxembourg just before its invasion of Belgium. Luxembourg, by the way, is a bilingual country, its inhabitants speaking French and German with equal ease.)

To get back to the bells, it wasn't long before I was making quite a good deal of money by installing bells and house telephones for friends and others and, since my expenses were practically nil, everything was clear profit—a condition which I have never been able to duplicate since that time.

My parents, who were well-to-do, looked upon this entire electrician business with sad misgivings, and my mother watched the proceedings with the same enthusiasm which a hen feels when she sees a duckling which she has hatched, take to the water.

Opposition, however, ever since childhood has never deterred me, and I kept on importing my electric supplies, from either Germany or France, until I acquired something of a reputation in my city.

My activity soon became known to the Mother Superior of the Carmelite Convent, a large institution occupying,

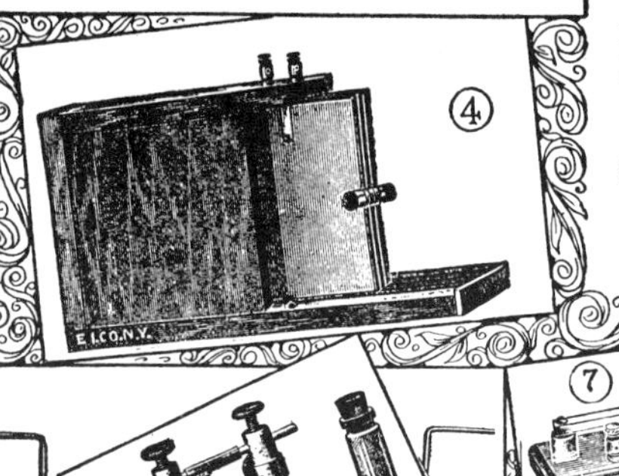

(The illustrations on this page are identified, by number, at the end of the article.)

E.I.Co. Days

apparatus and literature. In commemoration of 50 YEARS OF RADIO, the founder of that well-known company here gives a comprehensive story of its history which will prove of interest to old- and new-timers alike, telling as it does of the birth of a major American industry.

with my father's estate, a high plateau on the outskirts of the city of Luxembourg. I was a frequent visitor to this convent, and Mother Bodewing, the Superior, took a liking to me which was manifested by her presentation to me of various holy pictures, which I treasured highly. One day I was asked if I could provide an intercommunication system for the Carmelite nuns, from one cell to the other, so that the nuns could speak to each other when resting. The order of the Carmelites is quite rigorous; the nuns, for instance, use their coffins as beds, and wear very rough homespun sackcloth for their uniforms.

After a few consultations, a price was agreed upon, and a number of telephones and bells were to be installed in the convent. When the material arrived it was found, however, that although I was yet only 13 years old, I was still considered a man, and as such could not enter the premises of the convent. Indeed, Mother Bodewing had to get a special dispensation from Pope Leo XIII for me in order that I might enter the convent. In due time this permission arrived, and the telephone and bell system was installed. Upon the completion of the work, the Mother Superior presented me with a beautiful illuminated parchment in gold and many colors, testifying to the fact that I had, at the age of 13, installed a telephone and bell system to the satisfaction of the convent.

With the proceeds of the profits made from such telephone and bell installations I soon had assembled a creditable electrical laboratory. Since my parents did not encourage this sort of thing, I found it necessary to buy all supplies for the laboratory out of my own funds.

Not long afterwards, I learned about the "wireless" research of Marconi and, in a few days of hard work, had performed his experiments. With crude apparatus, I was able to send signals across the room with ease.

These wireless experiments, however, did not hold any great significance to me at that time, because I was much more interested in the mysteries of batteries. Indeed, most of my money was expended in making all sorts of primary batteries, and the designing of batteries which could be used for lighting homes. This was a necessity in those days, because, while the city already had an electric power plant, the district where we lived was not provided with service and, consequently, we had to be satisfied with smelly oil lamps.

It wasn't long, however, before I had electric light of my own, and had equipped our house with a number of lights also—all supplied by batteries of my own design. The stunt in those days was to find a good copper sulphate battery, a type known to give a steady stream of current without weakening, and with it charge a storage battery during the daytime; then by throwing a switch at night, light was provided. Of course, these lamps were of the miniature variety, and each gave about

(The illustrations on this page are identified, by number, at the end of the article.)

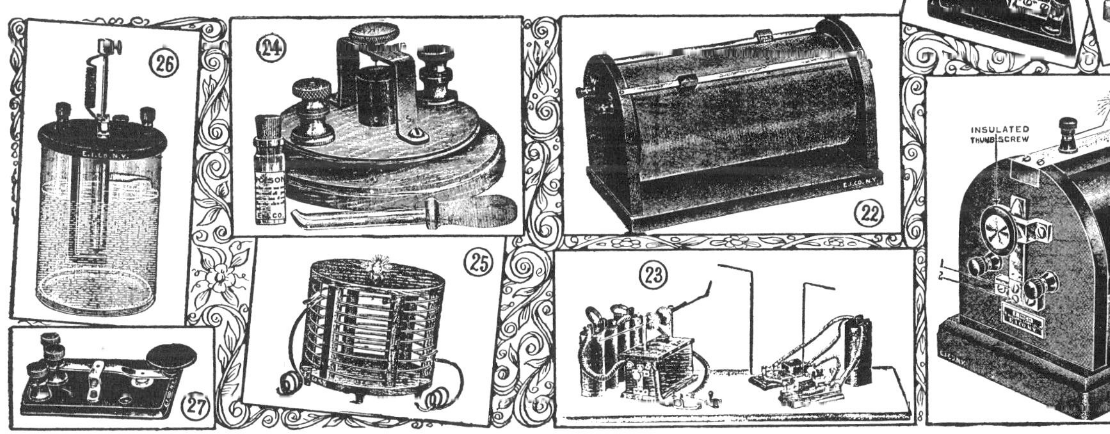

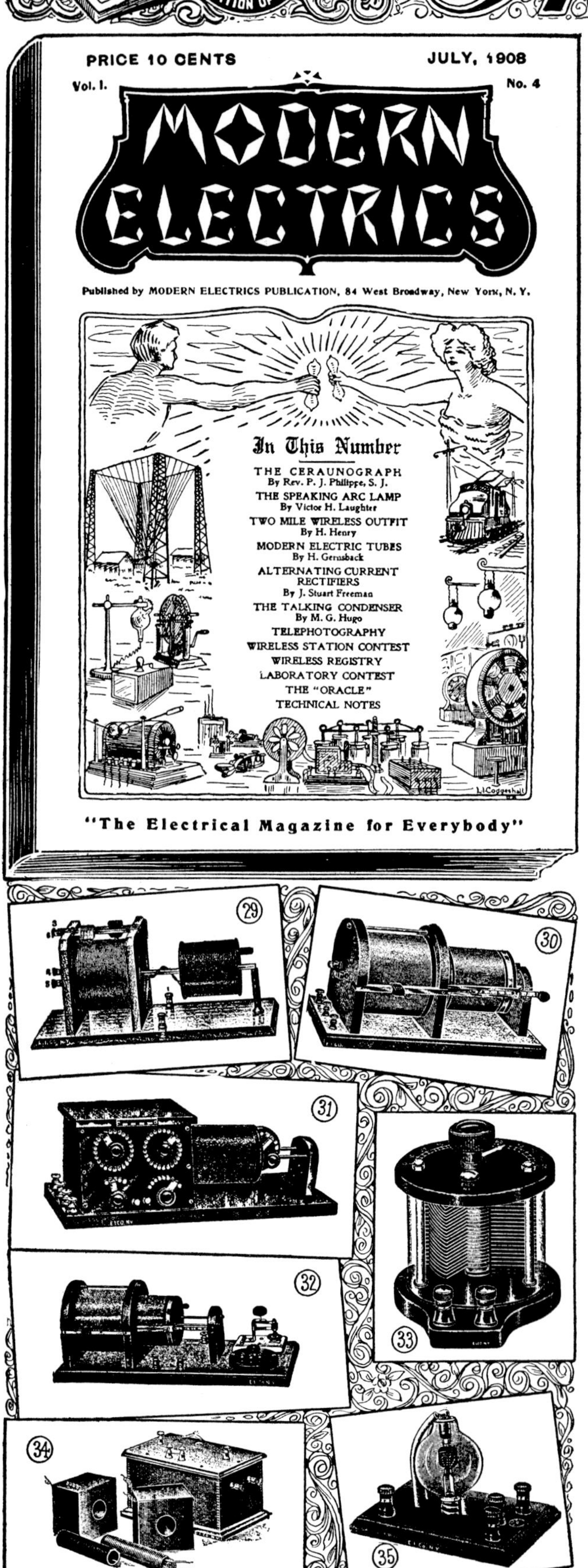
PRICE 10 CENTS JULY, 1908

Vol. I. No. 4

MODERN ELECTRICS

Published by MODERN ELECTRICS PUBLICATION, 84 West Broadway, New York, N. Y.

In This Number

THE CERAUNOGRAPH
By Rev. P. J. Philippe, S. J.
THE SPEAKING ARC LAMP
By Victor H. Laughter
TWO MILE WIRELESS OUTFIT
By H. Henry
MODERN ELECTRIC TUBES
By H. Gernsback
ALTERNATING CURRENT RECTIFIERS
By J. Stuart Freeman
THE TALKING CONDENSER
By M. G. Hugo
TELEPHOTOGRAPHY
WIRELESS STATION CONTEST
WIRELESS REGISTRY
LABORATORY CONTEST
THE "ORACLE"
TECHNICAL NOTES

"The Electrical Magazine for Everybody"

1 or 2 candlepower when the storage battery was freshly charged.

All this work was, of course, done after school, and the work continued during my studies in Brussels (Belgium), as well as in a technical college in Germany.

After the turn of the century—at the end of 1903—while still in my 'teens, I decided I wanted to come to America to exploit a battery invention, because I felt that there was no opportunity in Europe for this sort of thing. The battery which I had invented was a dry-cell fashioned like a storage battery, with alternating carbon and zinc plates in parallel.

The trouble with the invention was that the thin carbon plates had to be surrounded by a depolarizer and the arrangement was somewhat too fragile. This battery, which measured about 4 ins. square, and about 6 ins. high, nevertheless gave over 125 amperes on short-circuit and, under normal current consumption, 1½ volts. It was, however, found impractical in my rounds through American battery shops, because of its exceedingly high cost.

I modified the idea of my battery somewhat, and soon found a partner; so we started to manufacture ignition batteries for automobiles. This business succeeded, and would have been a success except that the depression of 1907 came along. A dry storage battery of my design which we were manufacturing at that time, although successful, could no longer be marketed during the slump, as most automobile plants had closed down. So, for the time being, the battery business ceased, and other activities took its place.

I had become acquainted with a young man named Louis A. Coggeshall, who had been a telegraph operator with the Erie Railroad Company, and I used to tell him about the wireless experiments I had made while still in knee breeches. He was much impressed with them; so, one day, I constructed a small wireless set, and he became rather enthusiastic about it. That evening we decided, right then and there, to go into business. We wrote down a number of names; but I was somewhat skeptical that we could succeed by selling just a wireless set, and suggested that it would be better to get out a small catalog and list in it a variety of electrical supplies—a number of which we could procure in Germany and France, from supply houses which I well knew from my former connections with them.

I suggested the name of the Electro Importing Company, which was later on to become famous as the first wireless amateur supply house in the world. We took quarters in a small office in the top of No. 32 Park Place, in downtown New York, just a block from where the Woolworth building now stands. No. 32 Park Place still remains as it was in those days—the building stands unaltered.

One of our first investments was in a small 1 inch advertisement which we inserted in the January 13, 1906, issue of the *Scientific American*. A similar ad was inserted in an issue of *Youth's Companion* at about that time. We were blissfully ignorant of the fact that these were, later on, to become historical as the first radio advertisements, whereby a radio set was offered for sale to the public for the first time in history.

The little set which we marketed was, of course, crude, but it worked! I had gone Marconi one better, and invented a new circuit, whereby only a single battery was used in a single circuit. Marconi always used a separate battery for the coherer circuit, and another for the decoherer circuit. I combined both circuits into one, thereby saving a battery. We applied for a patent on this circuit, but made the mistake of not employing a regular patent attorney since Mr. Coggeshall, my partner, did the regular application work himself. Many irregularities were found by the Patent Office, as to form and procedure and the application had to be sent back and forth so often that Mr. Coggeshall got discouraged, and we forgot about the whole thing.

(The illustrations on this page are identified, by number, at the end of the article.)

E.I.Co. Days

THE WORLD'S FIRST WIRELESS STORE · 69 W. BROADWAY · N.Y.

This wireless set which was called the *Telimco* (a coined name made up from the words *The Electro Importing Company*) was priced at $7.50. Orders began to come in pretty rapidly, and our only trouble was that we could not fill them fast enough. One of the difficulties was in the coherer filings, a mixture of silver and iron. Mr. Coggeshall had under him the "filings department," so-called, and it was his duty to take good U. S. dimes and file them with a coarse file, catching the filings in a small cardboard box. These were then put into a glass bottle and kept there until we mixed them with iron filings in the proper proportion.

As manufacturers of the first complete wireless set, we early had a brush with the law. It seems that some skeptical readers of the *Scientific American* and the *Youth's Companion* thought that a working $7.50 wireless set must, of necessity, be a fake; so they wrote letters to the Mayor's office in New York City demanding that we be investigated; the advertisements on their face were apparently a fraud. So, one morning, we were astonished to have a burly Irish policeman descend upon us, demanding in a bellowing gruff voice to be told just what sort of business was going on. Naturally, we assured him we were making a wireless set. The stalwart minion of the law bellowed in unmistakable terms that it was impossible that we could do such a thing for $7.50, and that, as a matter of fact, wireless was the most expensive thing going, and no set could be produced for less than $10,000!!

As our demonstration set was always hooked up, we asked him to take into his own hands the receiving part, which consisted of a board, not quite a foot square, on which there was a battery, coherer, decoherer and relay. The coherer was equipped with 2 mysterious looking antenna wires bent in the form of an "L". All the connecting wires were done, in the then-prevailing fashion, in beautiful spiral curlicues without which no electric apparatus was complete.

So there the cop stood in the middle of the room, balancing the receiving set in his ham-like hands. I betook myself to the end of the room where the transmitter (in the shape of a 1-in. spark-coil, transmitting oscillator balls with their antenna, four drycells and a transmitting key) was located. I then asked the policeman to state how many times he wanted me to ring the bell of the receiver which he had in his hand.

"Four!" he growled. I depressed the key 4 times and, the bell rang 4 times in the best wireless manner of the day. The cop was nonplussed, but still couldn't believe his senses. He then asked me to vary the signals to 3; then to 6; and finally to 12. Fortunately for us, that day the receiver performed miraculously well. (It did not do so on all occasions, due to some electrical disturbances in the building; usually the elevator motor when starting, set off our receiver.)

The policeman put down the receiving set. He was plainly astonished. Yet he had had his first real wireless demonstration—the bell had rung in his own hands without any connecting wires between the transmitter and the receiver!

Let it be stated here that a New York cop is not defeated so easily. His parting words, while he backed through the door, must be recorded here for history. They may not be grammatical, but they were to the point:

"I still think youse guys is a bunch of fakers. This ad here says that you are selling a *wireless* machine. Well, if you do, what are all them wires for?"

The wires to which the policeman referred were, of course, the antenna wires and the connecting wires; and, technically, we must bow to his superior judgment. He was right. Our wireless set *did* have wires, and all radio sets continue to have them to this day.

After the first furor which our set had created, there came the inevitable slump. With the coming of summer,

(*Continued on page 630*)

(The illustrations on this page are identified, by number, at the end of the article.)

OUR NEW 5 STORY BUILDING

MILEPOSTS IN

Ever since the invention of communications, man's mind has played with the possibility of sending pictures — moving, living pictures — winging their way through space. Even before science had solved the basic problems of dividing each image

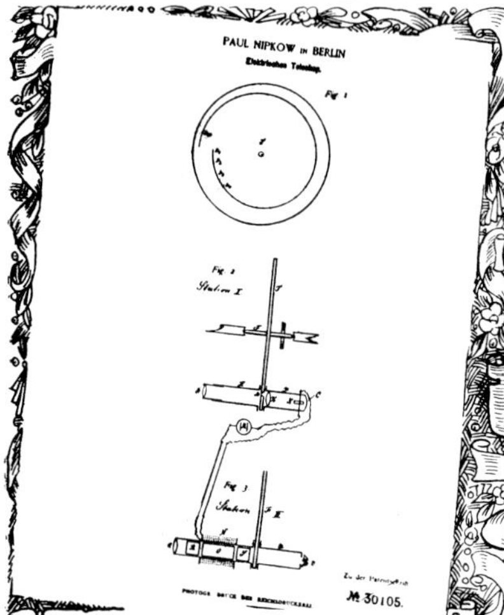

Reproduction of patent drawing of Nipkow's Scanning Disc - 1884.

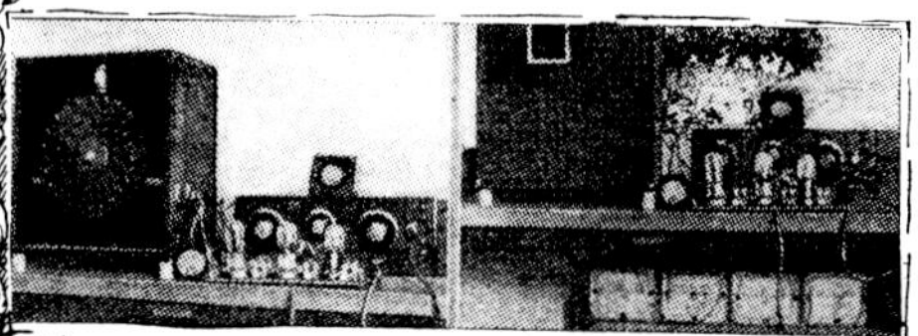

Home-constructors' early television receiver - about 1928.

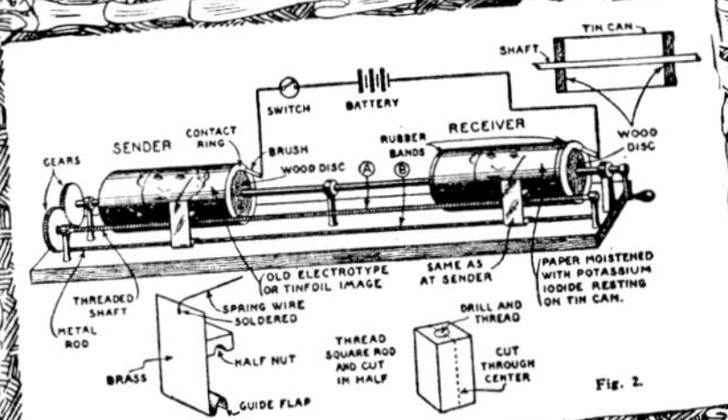

Early image receiver and transmitter - about 1927.

Television scanning transmitter by Insuline Corp. of America - about 1927.

Early "Baird" Commercial Television Receiver - 1927-'31.

Early "Freed-Eisemann" Commercial Television Receiver - 1927-'31.

Early "See-All" Commercial Television Receiver - 1927-'31.

Early "Continental Telev. Corp." Commercial Television Receiver 1927-'31.

Early "Jenkins" Commercial Television Receiver - 1927-31.

TELEVISION dates back far more than 50 years, although the first practical demonstration was not given until January 27, 1926, when John Logie Baird showed members of the Royal Institute in London his system reproducing images in black, white, and tones lying between.

As early as 1860, however, Nipkow of Germany, Carey of the United States, and Senlec, of France, were working out the first theories of this new art. The former, indeed, devised the perforated disc which bears his name, and which was the accepted means of television pick-up until about 1932, when the Zworykin *iconoscope*, an electronic pick-up, was proven to be much more efficient. This, however, did not solve the complete problem; that of reception remained.

FIRST PRINCIPLES

The art of sending moving images through the air also owes a debt to Lumière, Edison and other early developers of the motion picture, and to the numerous workers who devised halftone engraving, for it makes use of principles of which they took advantage. Television makes use of the limitations of the human eye; i.e., the eye's failure to separate a number of small, closely-spaced spots at a distance from it (as in halftone engravings) and its tendency to "see" an image a fraction of a second after the image has gone (as in motion pictures). Television takes advantage of such "resolving power" and "persistence of vision". It breaks (in modern 441-line scanning) an image into nearly 200,000 dots or "picture elements", only one of which is upon the screen at any given instant—the eye blends them into a complete image.

Even the younger television engineers can easily recall the days when television images were composed of no more than 516, 2,304 or 3,600 picture (or, more correctly, "image") elements (constituting 24-, 48- and 60-line scanning, respectively). Comparatively crude, they yet permitted the presentation of recognizable images and clearly foreshadowed the possibilities of television as a means of public and home entertainment.

Oddly enough, a successful television experiment was described in the public press only 7 years after the discovery of the properties of *selenium*—and 4 years *before* the simplest scanning disc

into a series of electrical impulses, human brains conceived, in a hazy way, some idea of how the end might be achieved. This article traces the progress of the art from its inception onward to its present state of high development.

was invented! Calling the system "diaphote", one H. E. Licks hoaxed the press of the United States by describing an imaginary experiment in which a selenium pick-up was used. It was not until 1884 that Paul Nipkow, of Germany, invented the *scanning disc*, using a spiral of pinholes to cover the objects at the transmitter and the image at the receiver in the process now known to the television world as "scanning."

THE NIPKOW DISC—BASIC SCANNING MEANS

For many years the *Nipkow disc*, or modifications thereof, was the basis of television experiment. C. Francis Jenkins placed his apertures in the surface of a cylinder, for example, while others retained the disc form, but substituted lenses for the pinholes, to pass more light. The spiral form was still retained in the early lens discs, however, and the spot of light, having to be large enough to cover both the first and last lenses in the series, was largely blocked by the uncut portions of the disc.

Using a unique form of reflecting lenses, angularly displaced, William Hoyt Peck was able to arrange these in a circle instead of a spiral, and so was enabled to concentrate his light beam to a small spot, all of which was reflected (save for the usual reflection and refraction losses) to the screen. This produced images of considerable size and brilliance. He put this apparatus into commercial use for bulletin transmission over wires late in 1937.

But the Nipkow disc was by no means the sole basis for television development, though it did mold thought in the art for many years. Other early ideas which must receive their meed of mention include the *mirror screw*, the *revolving prism* and the *cathode-ray tube*.

ENTERS THE CATHODE-RAY TUBE

Startling, isn't it, to think of the cathode-ray tube as an early system, when it has won almost universal adoption by the leading television laboratories of today? Yet Boris Rosing, a Russian professor, was conducting television experiments with cathode-ray tubes as early as 1907. Rosing used a mirror-drum scanner at the transmitting end, and had upon it a number of coils to generate synchronizing currents to cause electrical scanning in the cathode-ray tube at the receiving end. He was not the first, however, to make use of mirrors, for vibrating reflectors

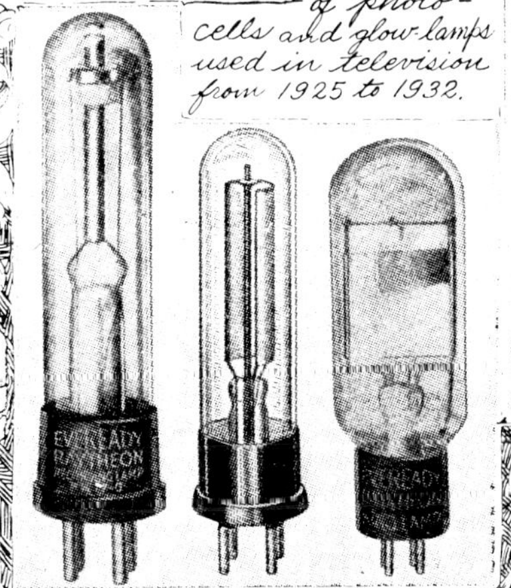

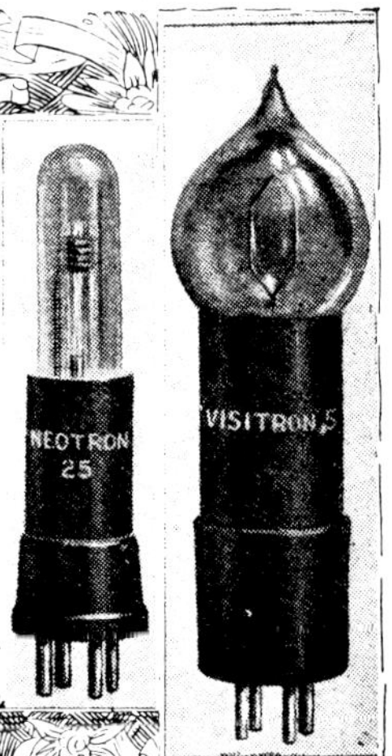

Early types of photo-cells and glow-lamps used in television from 1925 to 1932.

Scanning Drum televisor, by Jenkins, - about 1928

Jenkins-Moore television apparatus - 1925

Bell Labs' television scanner with lenses - about 1929.

Dr. Alexanderson's television projector; - about 1929.

Walter S. Gifford tests "telephone television" - 1927.

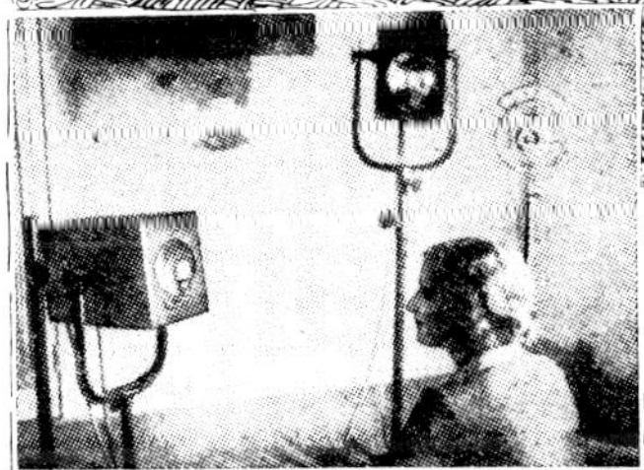

Early television camera or scanning arrangement - about 1929.

W1XAV, Boston, Pioneer Television broadcaster - about 1930.

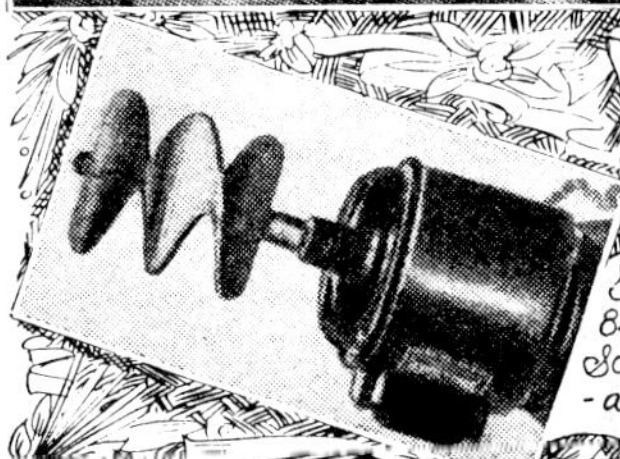

German (TE KA DE) 84-line "Mirror-Screw" scanner, - about 1930.

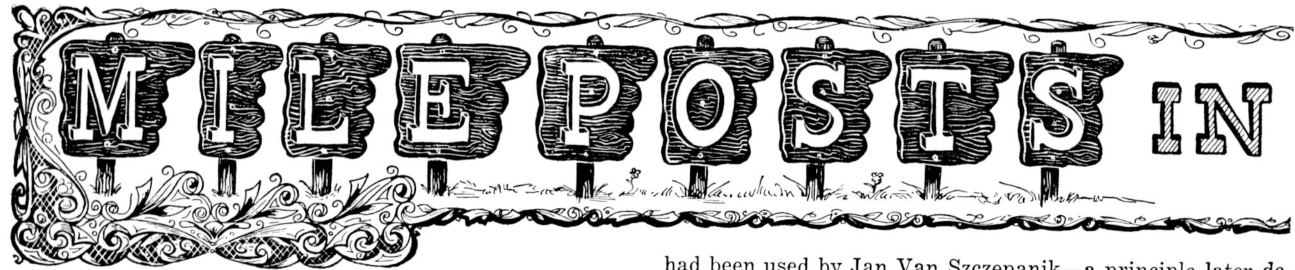

had been used by Jan Van Szczepanik—a principle later developed along far more efficient lines by Priess.

While scanning has occupied the attention of television engineers to a great degree, there has ever been the coexistent problem of light-beam modulation. Szczepanik used a method which was widely employed for a time—namely, directing the beam (at the receiver) through a small aperture, from which it was deflected for the more shaded picture elements. It was not until some years later that the *Kerr* (or Karolus) *cell* was used for the same purpose, but far more effectively, in connection with *polarized light.*

OUTSTANDING TELEVISION TRANSMISSIONS

Television broadcasts have been on and off the air at intervals since the late '20's. The first of any major importance were the broadcasts of television over WRNY, under the direction of Hugo Gernsback, who sent out signals on a broadcast wave, in 1928, using Pilot equipment. The next step was taken by Jenkins, who radiated televised "sight and sound" programs in connection with broadcasts originating at WGBS. Next, the Columbia Broadcasting System, through an ingenious system of dual modulation, sent sight and sound on a single wave.

The New York Telephone Company engineers, under the guidance of Dr. Ives, sent *natural-color images* during this period, but the images were so small and the cost of transmitting and receiving equipment so large that the experiments had little immediate practical value. Of greater interest to the average radio fan were the rather dim, pinkish images seen when peering through pinholes at the *neon-tube* plate.

THE NEON TUBE

In the late 1920's and early '30's, the *neon tube* gave the amateur television experimenter a means of conducting his work at a moderate cost. For a few dollars, he was enabled to buy a neon tube with an inch-square plate, to produce an image of similar size, scanned by a pinhole disc, which he viewed through a magnifying lens. Receivers of this sort were actually marketed to the public by the Jenkins and "American Baird" companies during that period.

U. A. Sanabria at his television pick-up transmitter about 1930.

Kerr (light-modulating) Cell, and system showing its application about 1931.

Early cathode-ray tube for television. Note absence of "election gun" about 1931.

Early Farnsworth cathode-ray television transmitting and receiving system—about 1931.

Early Zworykin cathode-ray television receiver—about 1931.

Early Western Television Corp. "commercial" television receiver—about 1931. W.T.C.

RCA-N.B.C. television receiver of about 1936. NBC

TELEVISION

Reception was considerably improved when the *neon "crater" tube* was produced. This had a tiny "crater" of brilliantly-glowing ionized gas, bright enough to project a picture a few inches square when used with a scanner in which lenses replaced the pinholes, to pass more light. Peck, using this principle, showed images about 2 by 3 feet.

SCREEN-SIZE TELEVISION

John Logie Baird, of England, E. F. W. Alexanderson and Ulysses A. Sanabria, of the United States, were early successes with large-image television. Sanabria used lens discs with a "neon arc" light and "interlaced scanning"; Baird a mirror drum, with a Kerr cell and arc light. The same sort of light source and modulator were used by Alexanderson.

Interlacing was an invention of no mean merit, as it decreased to a considerable degree, the flicker that was present in straight scanning. In simplified language, it consisted of scanning the odd-numbered lines completely, then the even-numbered lines completely to form a single image, thus reducing the "carry over" which the eye previously had to make as the dot leaped from the end of the last line of frame No. 1 to the start of the first line of frame No. 2.

A similar principle is used in the cathode-ray systems developed by RCA and Farnsworth.

CATHODE-RAY PICK-UP "CAMERAS"

Dr. Vladimir Zworykin and Philo T. Farnsworth, working independently, arrived at similar ends in the middle 1930's; these were the Zworykin *iconoscope* and the Farnsworth *image dissector*. Both provide electronic means of scanning an object, and translating into high-frequency electrical impulses the picture elements which result. The number of lines per image is 441 (roughly 200,000 picture elements per picture). In a 5 x 7 inch image (this is the equivalent of about an 85-screen halftone—almost as much detail as the cuts in this magazine which uses 100-screen). The images are scanned at the rate of 30 complete pictures or frames per second, using interlacing. (As we go to press we learn that Allen B. Dumont has developed a *new* television system. Read about it in the April, annual ELECTRONICS NUMBER, of *Radio-Craft.—Editor*) *(Continued on page 607)*

Philco television receiver, - about 1936 P.R&T.

Philco Camera Tube, for television scanning - 1937-'38

Television scanning of moving pictures. (Cameras project into iconoscopes) - 1937-'38 NBC.

The Philco Television Camera which scans at 441 lines - 1937-'38. P.R&T.

America's first mobile television transmitter, for outdoor events. 1937-'38. NBC

Iconoscope camera in action at N.B.C.- RCA's newest studio. - 1937-'38 NBC.

Newest (American) Dumont 12-inch cathode-ray receiving tube — 1937-'38.

Relative comparison of 25 line to 441 line image definition. (Note improvement as a greater number of lines are added.)

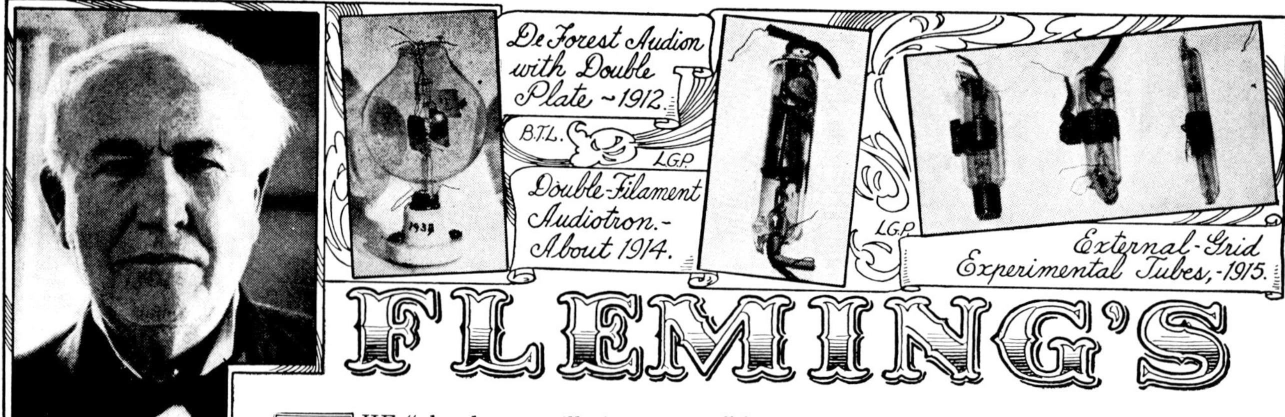

De Forest Audion with Double Plate ~1912.

Double-Filament Audiotron.- About 1914.

External-Grid Experimental Tubes,-1915.

Thomas A. Edison

FLEMING'S

THE "glow lamp oscillation detector" is the heart of your radio set—for that is what Dr. J. A. Fleming, of London, christened the vacuum tube when he adapted it to radio use.

Generally considered the pioneer in the tube field, Fleming in reality was merely one of a long succession of experimenters. It was his good fortune that thought was being devoted to wireless communication when he achieved his laboratory triumph.

Edisons Incandescent Lamp with Plate Element Responsible for Discovery of Edison Effect.

1725 TO 1905

The basis of the vacuum tube was laid down in 1725, when Dufay discovered that if one of two spheres was heated, a current-carrying path was formed between them. About 150 years later, Guthrie proved that if an iron ball, insulated, was brought to a red heat, it would still retain a negative electrical charge, but not a positive one, and 7 years afterward, in 1880, Elster and Geitel added a plate to an incandescent lamp. They noticed the *valve effect* by 1889!

Approximately half-way between the two latter dates (or in 1883), the Edison Effect—flow of current from heated filament to positively-charged electrode in a partially-evacuated tube—had been noted. Sir William Crookes had also conducted his experiments and shown that "corpuscles" (small particles of electricity, now called *electrons)* could be made to flow between 2 electrodes connected to a source of electricity (such as a Wimshurst machine), when the air between these electrodes was evacuated. The electrodes were sealed in either end of an elongated glass tube, now called a "Crookes' tube".

In 1883, as a result of his association with the Edison Electric Company of America, Fleming began his experiments; in 1897 Thompson published the results of a research into the conductivity of gases; and in 1905 Fleming (by that time with the Marconi Co., in England) patented his "glow lamp oscillation detector or oscillation valve", for he had found its valve (i.e., one-way) effect, and knew that it would function as a rectifier.

INTRODUCING "MR. GRID"

For a time, due to their insensitivity to weak impulses, these tubes or "valves" met with scientific interest but public apathy. Even when, in 1907, de Forest added the 3rd element, or grid, there was not much commercial activity in producing the audion for two reasons. The Marconi Company of England, which owned the Fleming valve patent, claimed an infringement since they asserted the de Forest tube to be a Fleming valve with a grid interposed. The basic patent was, therefore, Marconi's and de Forest could not make the tubes.

Also, since de Forest owned the audion patent, Marconi could not make the audion, and hence the audion situation reached a stalemate! The second reason was that the crystal detectors developed in the same year were far cheaper, and the results nearly as good as the audion.

The reason for the old audion's efficiency not excelling the crystal detector by proportions comparative to later years, was because so little was known of the precise phenomena which took place and the attendant need for very special precautions in manufacturing the tube. The early audion was very "soft" or gassy, and any plate voltage above (about) 30 volts would cause it to turn blue (gas ionization) and make the tube inoperative. Further additional plate voltage would result in complete destruction of the tube. With such low plate voltage plus the interference offered to the electron flow by the gas molecules, it was no wonder that the audion was only a mite better than the crystal detector. Hugo Gernsback's E. I. Co. was the first to sell audions to the public in 1911.

By this time the *diode* (2-element) and *triode* (3-element) tubes had undergone considerable analysis, and out of the analyses had come—and gone, thank goodness—many weird and fearsome designs. One of the most intriguing of these was the "external element" —either the plate, or the grid, as the case might be; numerous variations were tried and, in the final analysis, discarded, by many experimenters, including, Hugo Gernsback.

Fleming Valves or 2-Element Tubes,-1903-7

First de Forest Audion-about 1908

Replica of de Forest's Patent Application on Third Element Tube-1907.

Early de Forest Audion-1909.

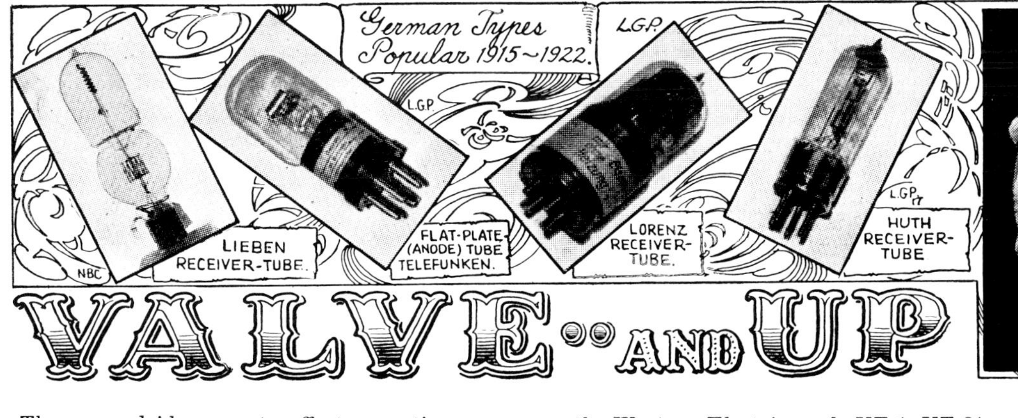

VALVE and UP

Sir Ambrose Fleming.

The general idea was to effect operation even through the glass wall of the tube; it worked, too, as real "old-timers" well recall!

HIGH-VACUUM TUBES

All these defects were eventually overcome as a result of Langmuir's researches into high-vacuum lamps (1912-1913) at the G. E. laboratories, and (almost simultaneously) Arnold's work in the same field for the Western Electric Company, a subsidiary of the American Telephone and Telegraph Company. While Langmuir's work was a result of incandescent lamp problems (although his discoveries affected the whole field of vacuum bulbs), Arnold's research was more directly centered about the audion—since de Forest had offered the tube to the A. T. & T. Co. in 1912, and they undertook to improve it for repeater (amplifier) operation for long-distance telephone communication.

In 1914, as a result of the A. T. & T. Co.'s work, the de Forest triodes were first used successfully in sending telephone messages from one side of the continent to the other and in the following year, in telephony from Arlington, Va., to both Paris and Honolulu.

The World War, of course, spurred activity in communications as well as in other fields, and while only gaseous or "soft" tubes had been available before, so-called "hard" tubes made especially for the Navy somehow got out to the public towards the end of and just after the war. Their superior operation and efficiency as compared to the old audion, resulted in a great demand, by amateurs, for good "hard"-vacuum tubes.

As a result of this demand for tubes by amateurs, we find numerous manufacturers entering this field in 1916-1919 and marketing their products—patents or no patents! Those first tubes will remain indelibly impressed on the memory of every old-timer. The "Moorhead", the "Electron-Relay", and the "Audiotron"—first with its double filament (in case one burned out), then later with a single filament, the various types of de Forest tubes—long tube-shaped (type T) and spherical, as well as the Western Electric-made VT-1, VT-21 and VT-2. When used in regenerative circuits with about 22½ volts for the plate—nothing that was ever made before for detection could touch them. Then, one day in 1920, the newly-formed Radio Corp. of America, with patent privileges retained from the old Marconi Company of America, began to make the Marconi VT tube and warn other companies that it was infringing on its Fleming patent.

THE "UV" SERIES

In 1921, RCA released the UV 200 (detector) and UV 201 (amplifier), both triodes with brass shells known as the UV base, and incorporating a filament that required one ampere at 5 volts for operation. Previously, during July, 1920, the General Electric Company (which was manufacturing the tubes for RCA) made arrangements with the American Telephone and Telegraph Company, owners of the de Forest 3-element audion patent, so that by an exchange of radio patent privileges each was allowed to make tubes without fear of litigation. Shortly afterwards, the UV 201A (thoriated filament) tube was developed and released, a far better tube than the old 201 in many respects, especially since the filament only consumed one-fourth of an ampere and hence materially reduced the drain on the "A" battery.

Many of the early tubes had oxide-coated tungsten filaments, and operated at low brilliancy—such were, for example, the WD11's, the 200 and 201. Meanwhile the laboratories had been experimenting with thorium, and found it to be an efficient emitter of electrons. Consequently, in 1922, tubes with filaments both coated and impregnated with thorium (200A and 201A) came into general use.

CONTEMPORARY DEVELOPMENTS

A digression here may not be amiss. These early tubes were high in cost as compared with tubes of today; they were not especially durable, and were easily damaged by being operated at too great a filament potential. Also, despite the ever-expanding facilities for tube

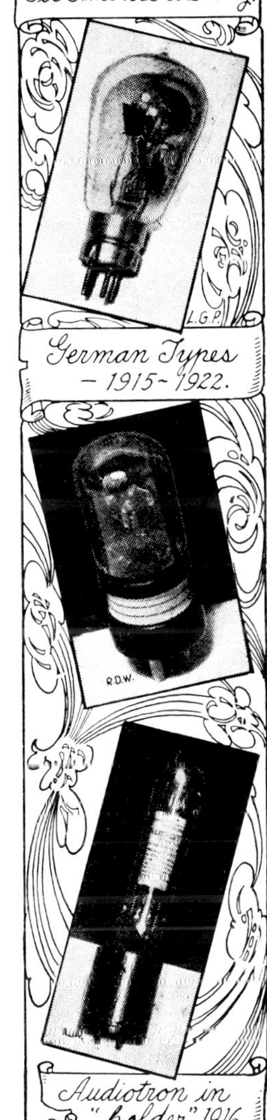

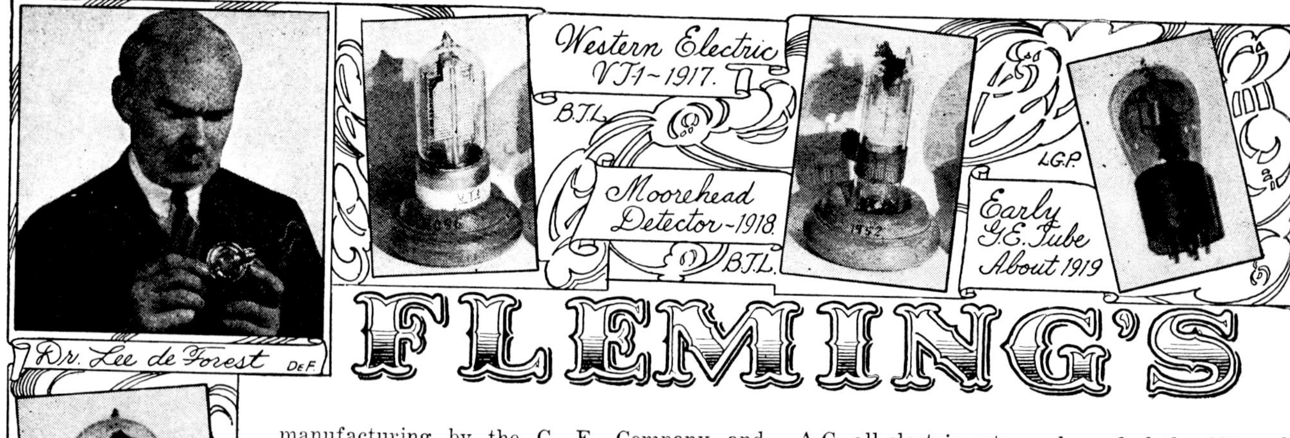

Bartley Tube—about 1919.

Electron Relay (Moorehead)-1920-'22.

manufacturing by the G. E. Company and Westinghouse (which were also manufacturing the tubes for RCA by this time), the supply could not keep up with the tremendous demand which was caused by amazing sales of receivers and parts as a result of the growth of broadcasting. Therefore, there sprang into being a new industry—*the rebuilding of radio tubes;* new filaments inserted for a dollar or two. Usually the repaired tubes had imperfect vacuum, but no one seemed to mind. At the same time, *tube reactivators* were put on sale for public use. These devices served to "boil" new thorium out of the wire at a high filament temperature, then "bake" it fast to the surface of the filament at a temperature only slightly higher than that used for normal operation. No plate potential was applied during this process.

The next advance was the low-current tube —the 199 (1922), which used only 60 milliamperes of filament current. This was important to fans who had to buy drycell "A" batteries or have storage "A's" recharged. After that, we find the trend going back to "soft" or gassy tubes for detection purposes. The 200A is an example, as well as the Donle sodium-vapor detector tube made by Connecticut Tel. & Tel. Co.

1926—POWER TUBES AND RECTIFIERS

In 1926, the early power tubes, such as the 120, the 112 and the 171 came into use. These afforded greater undistorted output and hence better reproduction than was formerly possible. The 210 and 250, usually used with a lighting-line filament and plate supply, soon followed.

The *"B"-eliminator* became popular in 1926, and as a rectifier the 80-type (filament) tube and Raytheon BH (helium-filamentless) or "cold" tube shared honors, although the latter tube is now practically unheard-of. All these types up to this date were known as *direct-heater* tubes.

1927—CATHODE-TYPE "INDIRECT-HEATER" TUBES

Then came 1927, with increased interest in A.C. all-electric sets, and we find the 226 and 227 tubes released by the tube manufacturers.

The 226 is a tube whose filament is slow to respond to the cyclic changes of alternating current, and, therefore, could be connected across an A.C. supply of 1.5 volts without causing hum.

The 227 employs a *cathode,* and is known as an *indirectly-heated* tube and for that reason was perfect for A.C. heater operation. Soon other A.C. tube models were produced, and the battery-operated tubes were pushed out of the radio picture, save for use in locations where electric mains were not available.

1928—SCREEN-GRID TUBES

In 1928, the *screen-grid tube* was released, type 224 for A.C. sets and 222 for D.C. sets. These tubes had tremendous amplifying power (as compared to old triodes) as well as inherently lower internal tube capacity.

The screen-grid tube did away with the necessity for neutralizing R.F. stages, and resulted in a more stable and sensitive receiver. In fact the amplification of this screen-grid tube was so great that local stations were tuned-in sometimes interfering with each other, sometimes so that they could not be "turned-down" soft enough. Special dual volume control circuits were necessary, as well as local-distance switches to reduce the antenna signal to the set on local stations.

The defect was known as "cross-modulation" and worried engineers no little, until 1931 when *variable-mu* tubes such as the '35 and '51 (by this time it had become the custom to use an apostrophe ['], later dropped, in place of the first of a tube's 3 code numbers) were released along with a *power pentode*—the '47 tube.

PENTODES

The new power tube was more sensitive and could be satisfactorily driven by the output of an average detector stage and still give higher power output.

In 1933, *multi-purpose tubes* were released such as the 2A7, 58, 2B7, 55, etc., to be followed shortly by similar tubes *but with heaters re-*

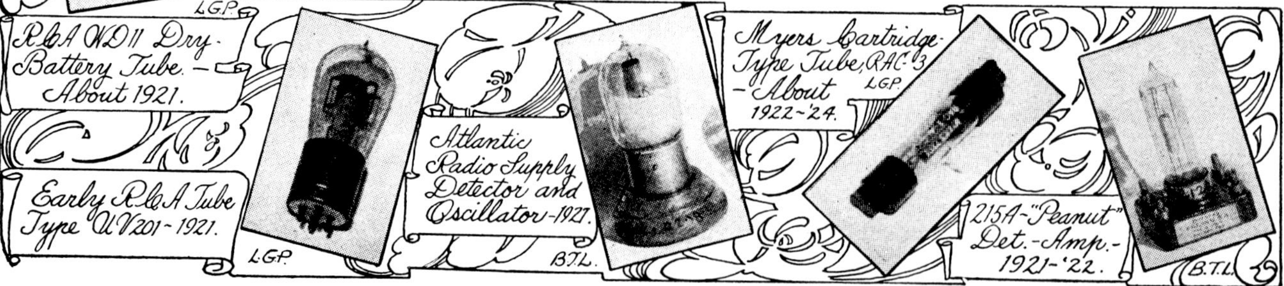

First "Peanut" Tube. With External Plate About 1922.

RCA UV201A Thoriated-Filament Tube ~1922.

De Forest DV3 Triode About 1923.

Dr. Vladimir Zworykin

VALVE .. AND UP

quiring 6.3 volts instead of 2.5 volts. The 6A7, 78, 6B7, etc., are examples. Also, in this year, the *acorn tubes* were developed as a result of research in the ultra-highfrequency field.

1935—METAL TUBES

The final stage in the development of tubes, to date, is the now famous *metal tubes*, and the *secondary emission* tubes. In theory, the metal tubes are a marked improvement over the glass tubes, since they are smaller, entirely shielded and, therefore, better for efficient, stable circuit operation. The "secondary emission" type of tube, in its most advanced design, is perhaps best represented by the *electron multiplier* introduced in the winter of 1935-'36 (See *Radio-Craft*, Jan. 1936, pg. 391.); its design makes possible a gain of several millions (i.e., amplification about equivalent to that of 6 or more ordinary tubes). The "electron multiplier" tube makes possible heretofore unattainable results in the field of television. All in all, the vacuum tube development from Fleming's early diode has been more than satisfactory, and in no small measure responsible for the great improvement in the operation of present-day radio receivers.

CONTEMPORARY DEVELOPMENTS

All vacuum tube developments, however, were not along "receiving tube" lines only. In the early part of the 1930's we find tube manufacturers considerably interested in various other applications of "electronics", as a consequence of which numerous strange types and special-purpose tubes have been developed. Their continuous research into the problems relating to tube operation and production made the evolution of special tubes designed for a specific though relatively foreign application, comparatively simple.

Specific types, with which we are now all familiar, are the various types of photocells including the caesium and the miniature RCA types 922-923; the cathode-ray, of which we now have various sizes ranging from a 1-inch in diameter screen to 10 inches; the electron-ray tuning indicator tube; the "iconoscope", a special tube for television pick-up purposes; the beam-power tube, used essentially in high-power amplifiers for obtaining tremendous power outputs; improved-efficiency transmitting tubes, by using graphite or carbon anodes for reducing plate losses due to the tremendous heating of this element when operated at maximum ratings. Also, concerning transmitting tubes, the improved construction of present types enables more stable and improved operation on higher frequencies.

Apace with these developments, we find improvements made in rectifier tubes—in some, the inclusion of mercury vapor which ionizes when the filament becomes heated and thus aids the conduction or flow of electrons between plate and filament. In addition, special types of thyratrons have been developed for inverter equipment, the release of which has done much towards solving many special problems in industrial fields.

It is perhaps best that mention be made here of special tubes *still in the laboratory and in the process of being developed*, the release of these tubes pending the solution of some problems which retard their efficiency and hence application in the radio and electronic fields. These tubes are of the "cold-cathode" variety, no filament or heater being employed although, it must be emphasized that no immediate promise for their future release to the public is in sight.

Due to lack of space we have not been able to discuss *transmitter-tube* developments. In general, though, they have paralleled the development of receiving tubes; and, in fact, almost all the low-power transmitting tubes, for many years, were taken from the field of receiving-type tubes. *Public address* and *electronics*, two newer applicational fields, tell the same story; only within the last 3 or 4 years has there been any evidence of other than regular radio receiving tubes being applied to these fields. But the march is on! Today, it is estimated that there is an approximate total (radio receiving and transmitting, public address, and electronics general and special-service) of about 500 types of tubes; tomorrow?—!

Schickerling Novel-Construction Triode—About 1924.

Conn. Tel. & Tel. Sodium Detector Tube—About 1924

Helium Rectifier; 250 ma. Output—for "A" Eliminators—About 1925-6.

B.T.L.

Newest and Possibly Future Radio Tube. The Electron Multiplier—1938.

Latest All-Metal Radio Tubes 1935.

G.E.

R.D.W.

Kellogg (Also Mc Cullough). A.C.-Filament Tube—About 1926

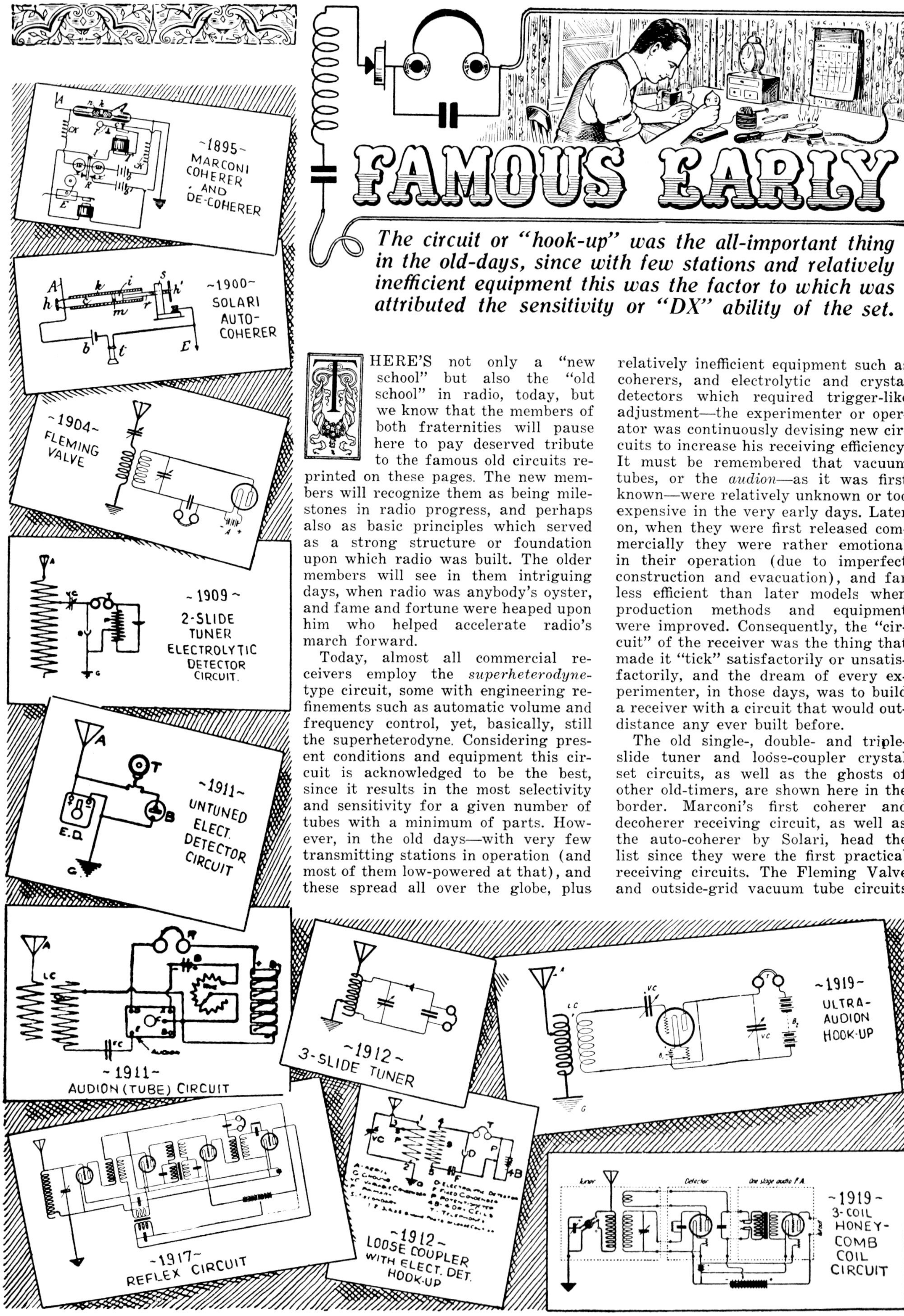

FAMOUS EARLY

The circuit or "hook-up" was the all-important thing in the old-days, since with few stations and relatively inefficient equipment this was the factor to which was attributed the sensitivity or "DX" ability of the set.

THERE'S not only a "new school" but also the "old school" in radio, today, but we know that the members of both fraternities will pause here to pay deserved tribute to the famous old circuits reprinted on these pages. The new members will recognize them as being milestones in radio progress, and perhaps also as basic principles which served as a strong structure or foundation upon which radio was built. The older members will see in them intriguing days, when radio was anybody's oyster, and fame and fortune were heaped upon him who helped accelerate radio's march forward.

Today, almost all commercial receivers employ the *superheterodyne*-type circuit, some with engineering refinements such as automatic volume and frequency control, yet, basically, still the superheterodyne. Considering present conditions and equipment this circuit is acknowledged to be the best, since it results in the most selectivity and sensitivity for a given number of tubes with a minimum of parts. However, in the old days—with very few transmitting stations in operation (and most of them low-powered at that), and these spread all over the globe, plus relatively inefficient equipment such as coherers, and electrolytic and crystal detectors which required trigger-like adjustment—the experimenter or operator was continuously devising new circuits to increase his receiving efficiency. It must be remembered that vacuum tubes, or the *audion*—as it was first known—were relatively unknown or too expensive in the very early days. Later on, when they were first released commercially they were rather emotional in their operation (due to imperfect construction and evacuation), and far less efficient than later models when production methods and equipment were improved. Consequently, the "circuit" of the receiver was the thing that made it "tick" satisfactorily or unsatisfactorily, and the dream of every experimenter, in those days, was to build a receiver with a circuit that would outdistance any ever built before.

The old single-, double- and triple-slide tuner and loose-coupler crystal set circuits, as well as the ghosts of other old-timers, are shown here in the border. Marconi's first coherer and decoherer receiving circuit, as well as the auto-coherer by Solari, head the list since they were the first practical receiving circuits. The Fleming Valve and outside-grid vacuum tube circuits

RADIO CIRCUITS

Those which had the greatest appeal, and hence gained everlasting fame, are herewith reprinted. Incidentally, many still form the basic groundwork upon which the more complicated, modern radio receivers are designed.

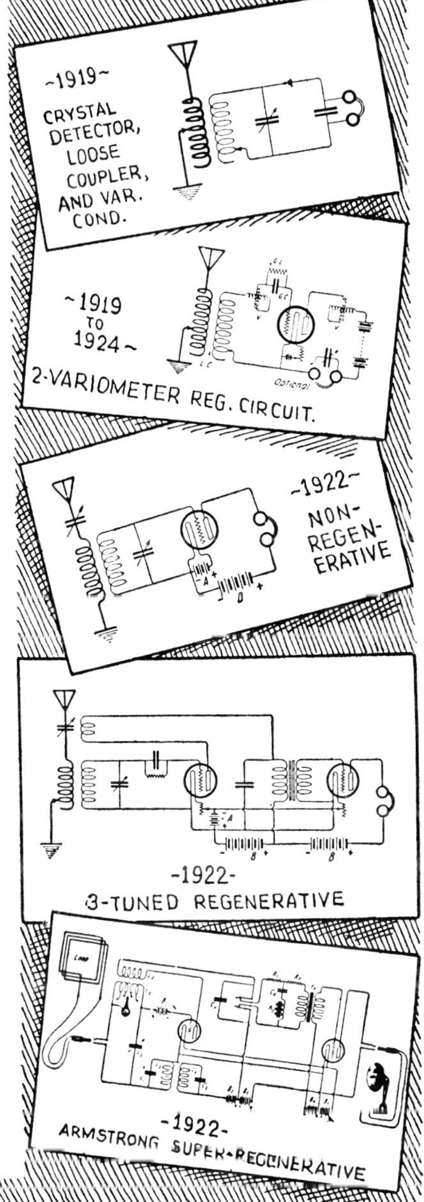

are also shown, to illustrate the first vacuum-tube circuits that were ever conceived. These were followed by the more popular de Forest Ultra-Audion and Single-Circuit *regenerative* circuits, although there were dozens of combination patch-type circuits in the interim (using tubes *and* crystals).

Later on, multiple tuned circuits became popular, using honeycomb coils to cover a wide-range of tuning frequencies. About the same time we find that variometer and variocoupler circuits, for tuning both plate and grid circuits and for regeneration control, became popular. These were followed by the still internationally famous Reinartz circuit, the Super-regenerative and Flewelling circuits, Hazeltine's Neutrodyne, Armstrong's superheterodyne with its multitudinous dials (when it first came out), the Reflex—and Grimes' variation called the Inverse-Reflex, Cockaday's Four-Circuit Tuner, the Ultradyne—a different type of superheterodyne created by Lacault.

Oh, yes,—we mustn't forget Colpitts, Hartley and Meissner whose circuits now form the basis for most transmitters in amateur and commercial work, as well as simple short-wave regenerative sets. And, finally, the first batteryless receiving circuits, developed in the early days of broadcasting when the crying public had its fill of crystals, storage batteries, eliminators, trickle chargers and other wasteful and devastating components.

While pet circuits are still the hobby of some experimenters—especially the short-wave fan who always strives to accomplish much with limited parts and finances—we do not hear or see so much of stunt circuits. The reason for this may be attributed to the development of screen-grid tubes, from the old 224A (or '24A) to present-day high-efficiency types. The inherent characteristics of modern tubes are such as to provide tremendous amplification, great selectivity (due to variable-mu characteristics) and inherent circuit stability due to the extremely low internal capacities of such tubes. Consequently, after the screen-grid tubes were first released, the neutrodyne or other self-neutralizing circuits became obsolete, and simple T.R.F. (tuned-radio-frequency) circuits without frills or innovations became the stand-by. While the T.R.F. circuit is not now as popular as the superheterodyne, it is still employed in a great many commercial receivers because of its simplicity and low production cost.

Today we have the most sensitive and

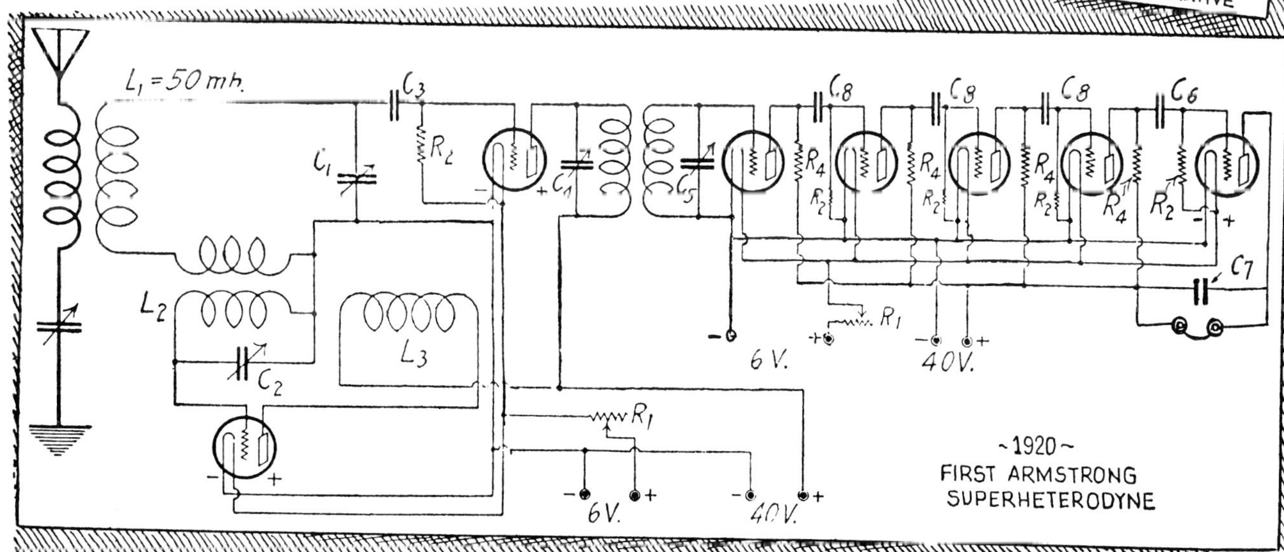

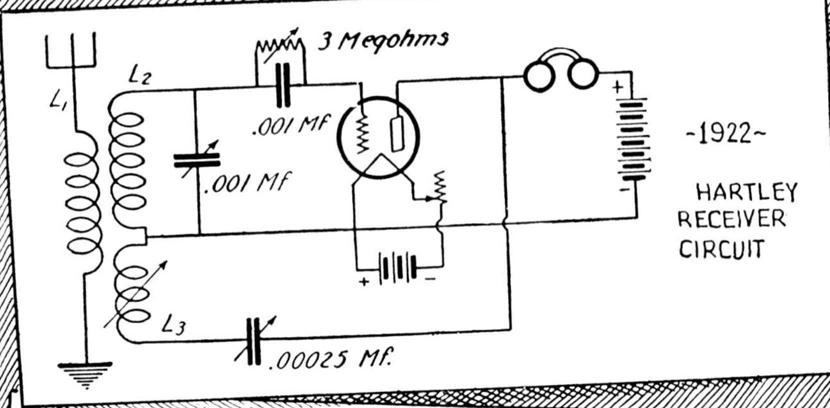

-1922-
HARTLEY RECEIVER CIRCUIT

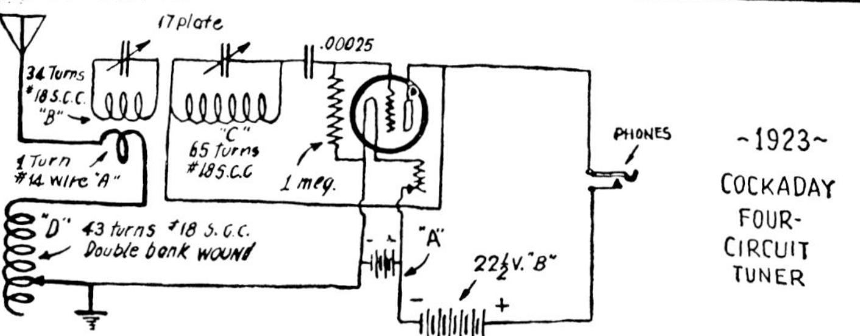

~1923~
COCKADAY FOUR-CIRCUIT TUNER

selective receivers ever developed, because of the superhet. circuit and with this circuit are combined many new, novel refinements, comparatively recently discovered. Specifically, these refinements are "automatic volume control" (A.V.C.), "noise suppression control" (N.S.C.), "automatic frequency control" (A.F.C.)—which permits a circuit to adjust itself to proper resonance when the operator tunes a set too casually—and, finally, "automatic tone control" in quite a few of the better receivers. These circuits are not shown in these pages for the very obvious reason that they have been discussed and shown very frequently and in very recent issues of this magazine. Perhaps some day they, too, will be considered obsolete, useless, in much the same light that we look upon the circuits of "old".

While none of the circuits are given here for the trickle chargers and "A" and "B" eliminators of that era, despite their role of importance in the progress of radio receivers, it is perhaps best that they are omitted for the simple reason that they were more infamous than famous. Undoubtedly, most of the

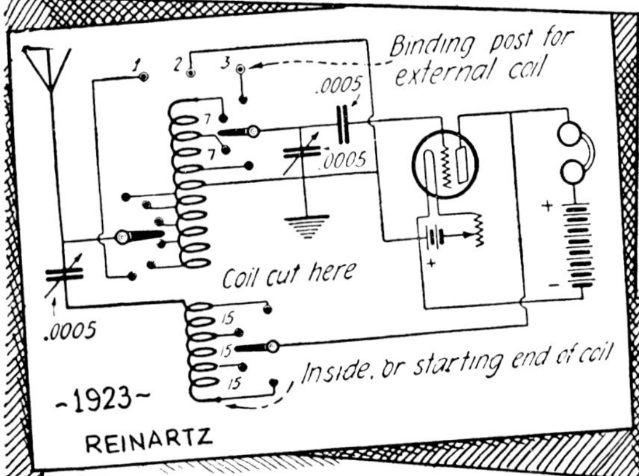

-1923-
REINARTZ

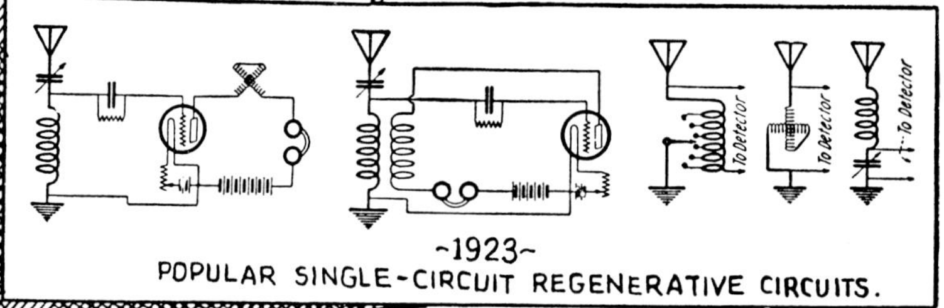

~1923~
POPULAR SINGLE-CIRCUIT REGENERATIVE CIRCUITS.

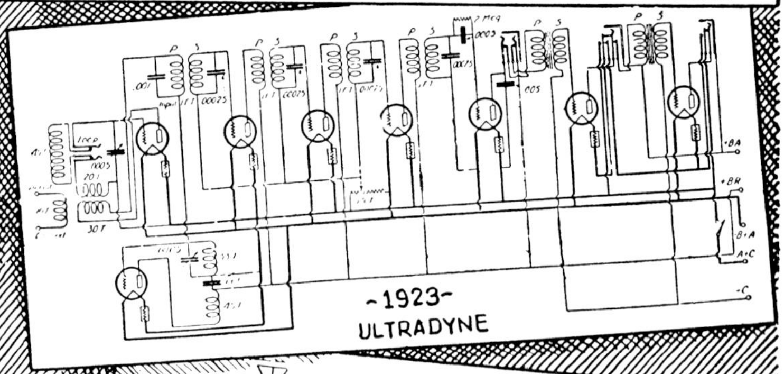

~1923~
ULTRADYNE

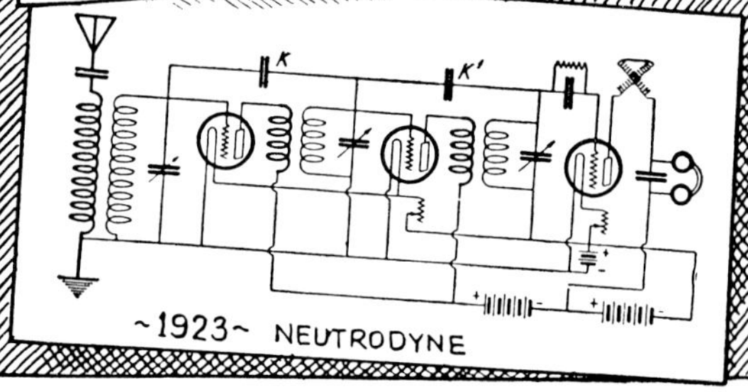

~1923~ NEUTRODYNE

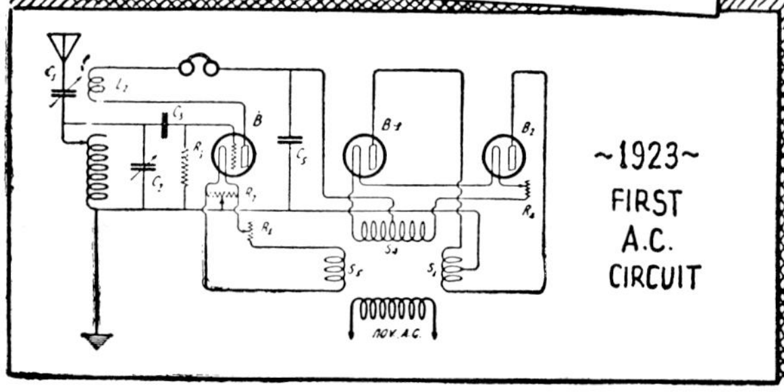

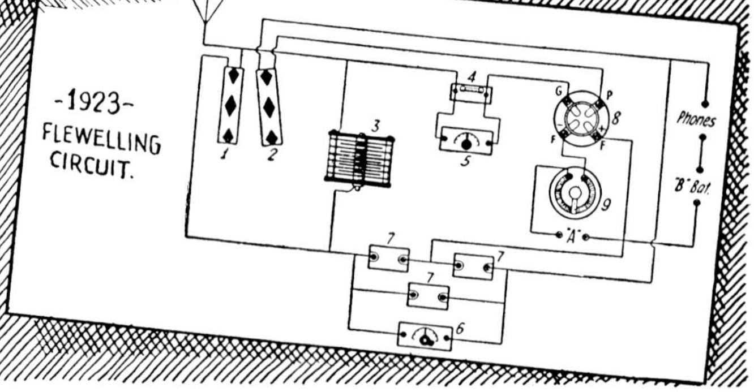

-1923-
FLEWELLING CIRCUIT.

~1923~
FIRST A.C. CIRCUIT

RADIO CIRCUITS

readers of these pages will remember them more for the nuisance which they created than for their historical role in radio. Concerning the first really "A.C." (alternating current) electric sets which came out shortly afterwards (1927), those employing the types 226 ('26) and 227 ('27) tubes, their era was so short and the circuits so easily available in various manuals that it is considered unwise to reprint them. After all, the real old-timers—such as the aforementioned—are the ones which made history, and which gave the experimenter of those days the real thrill.

Of course, the story would be incomplete if we were to leave out the circuit of the newest modern receiver, with all refinements such as A.V.C., A.F.C., pushbutton tuning, visual tuning indicator, etc., etc. Radio circuits have traveled a long way in 50 years, and the final diagram (the RCA model 811K, shown below) tells the story better with one illustration than could mere words.

Here they are,—**all On Parade!** How many of them do you remember, *and how many* have you tried out in your time? What fond reminiscences they bring back as we look them over!

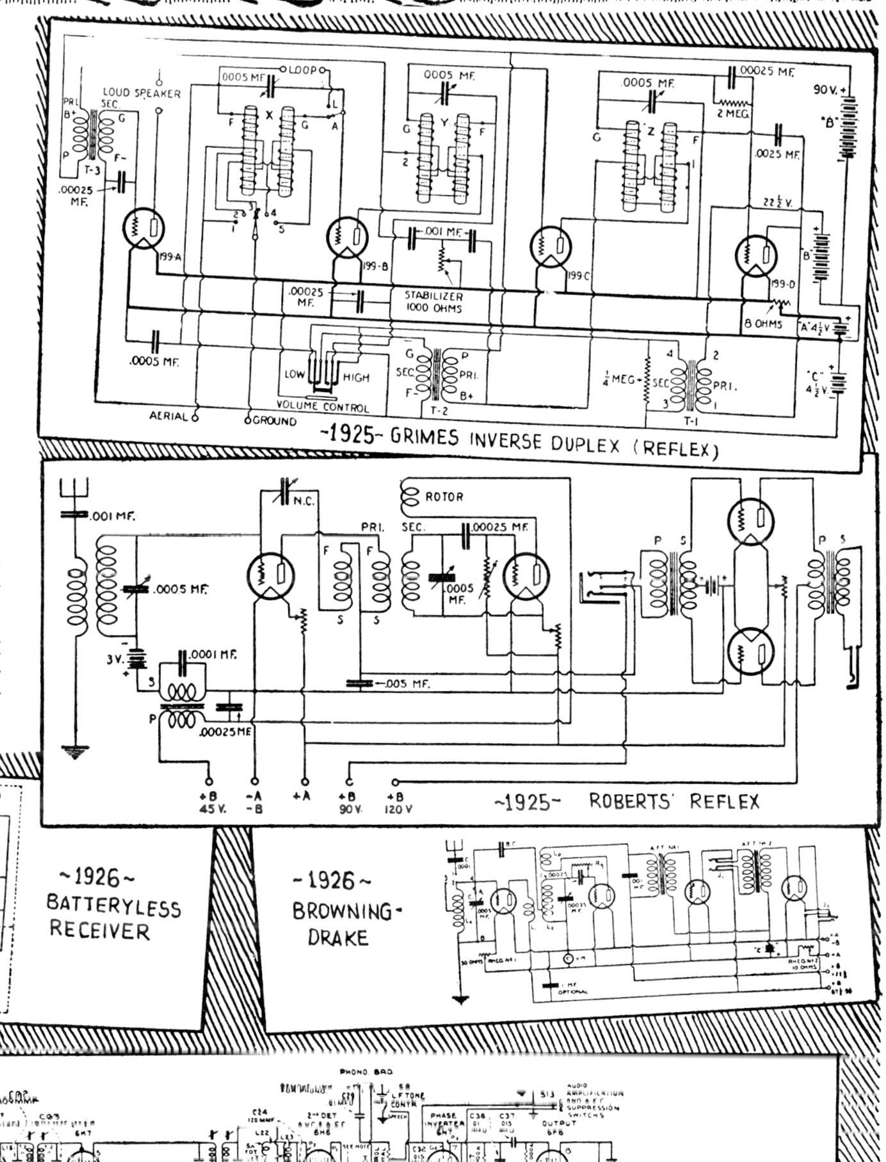

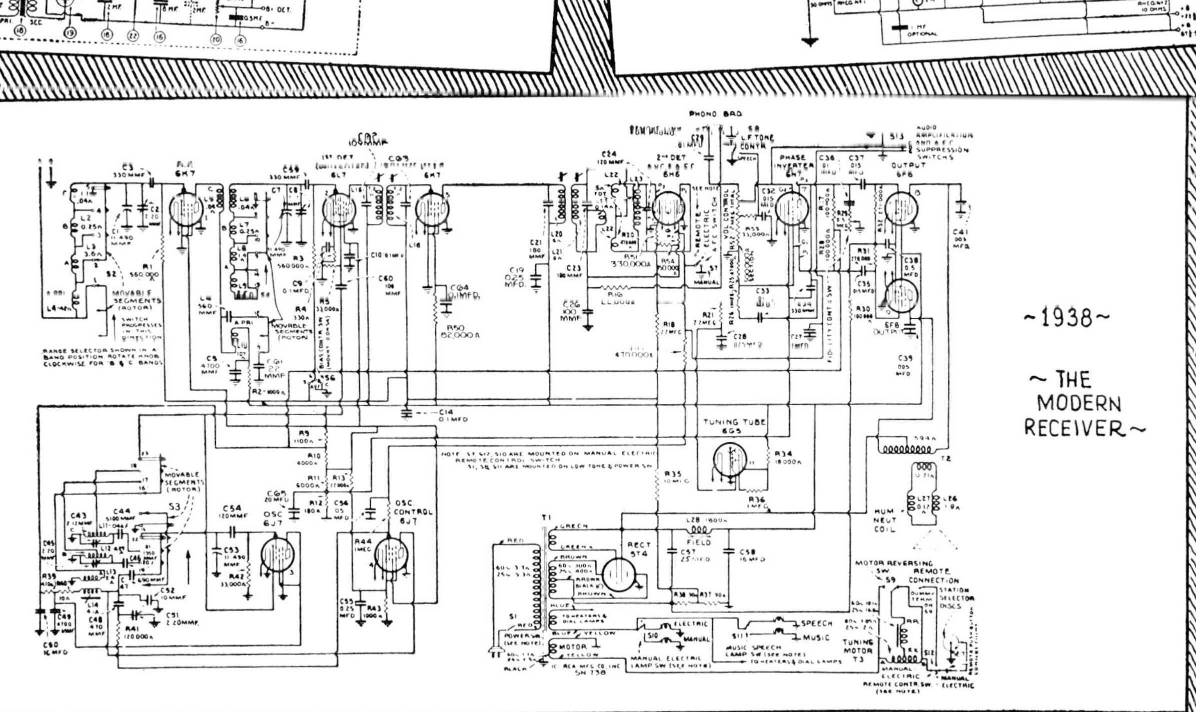

SOME OLD-TIME

The wireless stations recalled in this article will undoubtedly bring up fond memories to every old-time operator. The names of the "ops" who operated these stations in the old days may no longer be remembered but the service they

OPTICAL TELEGRAPH COMMUNICATOR ABOUT 1800.

An Early Marconi Station, about 1896 RCA

Glace Bay (GB) Transmitting Room, about 1904 RCA

BEFORE wireless came about, when a ship left land it was never heard from again until the day it returned to the same port, unless some other ship brought in news or tidings, or perhaps even some mail exchanged in mid-ocean. When wireless came along that picture was changed, since its first adaptation was for ship-to-shore communication although some of the first stations were intended to replace the telegraph and trans-Atlantic cable in the art of communication.

The *first* commercial wireless station was, naturally, Marconi's, and was erected in November, 1897, at Alum Bay on the Isle of Wight, England. The station employed an antenna consisting of wires stretched between masts that were 120 feet high, and many messages were sent from this station to an experimental station located on a steamer. These tests were designed to show the feasibility of ship-to-shore communication, and the first record of a paid-for message by wireless was made in June, 1898, when Lord Kelvin came to this station and sent a message, for which he paid, to another station at Bournemouth, approximately 14 miles away.

From then on, ship and shore stations were installed rapidly, and we find Marconi's company interested in spanning the Atlantic. With this in mind, he had already installed (in 1901) a powerful station at Poldhu, Cornwall, England. It employed for an antenna system 20 masts, each 200 feet tall, arranged in a circle 200 feet in diameter. After it had already proven that it could reach ships far out at sea, Marconi then traveled to St. Johns, Newfoundland, and after some experimental work was able to hear the letter "S" (3 dots: ". . . ") transmitted by Poldhu. Other difficulties arose, and it wasn't until 1902 that Marconi was able to return to the North American continent to erect his station for trans-Atlantic communication. When he did, he selected Glace Bay, Cape Breton, Nova Scotia, and on December the 17th of that year the Governor-General of Canada and the King of England exchanged greetings through the facilities of these two stations.

These were the forerunners of commercial ship and shore stations to which wireless operators were destined to listen eagerly for communications, press, instructions, weather reports, etc. In the United States, between 1904 and 1908, the United Fruit Company found it advisable to erect a number of shore stations so that its fleet of steamers traveling to southern climes

POZ or Nauen, Germany Early Radio Equipment. About 1905 S.P.S.

Receiving Room at Glace Bay Station (MARCONI AT LEFT) RCA

De Forest Radiophone Station on the U.S.S. "Connecticut," about 1908 RCA

RADIO STATIONS

rendered and the role that these stations played in the days when they constituted the single meagre ethereal thread which permitted contact between ships and distant shores will always remain alive and vivid to all seagoing operators.

MODEL OF MARCONI'S FIRST WIRELESS TRANSMITTER-ABOUT 1896.

could be contacted at will. These early United Fruit stations were located at Limon, Costa Rica and Bocas del Toro, Panama. Before 1908 this company had added more shore stations located at Nicaragua, Cuba, Guatemala, Louisiana, and Swan Island in the Caribbean Sea. In 1904, de Forest installed 5 large Navy shore stations at the following locations: Pensacola, Key West, Guantanamo, Colon (Canal Zone) and San Juan (Porto Rico).

Arlington (NAA) didn't go on the air until February, 1913, although the installation was started in 1909. Its first signals were made by a 100 kw. spark transmitter of Fessenden design, but this was replaced shortly afterwards by a 30 kw. "arc" transmitter which was much more efficient. From then on, arc transmitters were preferred by the Navy, and Arlington results were the criterion. The Eiffel Tower station (old "FL"), in Paris, began its transmissions in the latter part of 1912, and Time Signals from this station were transmitted regularly.

Marriott's Early Radiophone Transmitter at Manhattan Beach (old DF) Station, N.Y. 1908 M.

Shack and Antenna for Key West Station, about 1905 RCA

LONG-WAVE STATIONS

An interesting point, with which only real old-timers would be familiar, is that many of these aforementioned stations operated on extremely low frequencies or very high wavelengths, although the best results are obtained today in the opposite direction, i.e., ultra-highfrequencies or low wavelengths.

Arlington, or NAA as the veteran wireless operator would refer to this station, for instance, operated on a wavelength of 6,000 meters. Eiffel Tower, or FL, operated on a wavelength of 10,000 meters; Nauen (Germany), or POZ, operated on 12,600 meters, and NSS (Annapolis, Md.) operated on a wavelength of 16,900 meters or approximately 18 kilocycles —a frequency now considered on the fringe of the audio

(Continued on page 619)

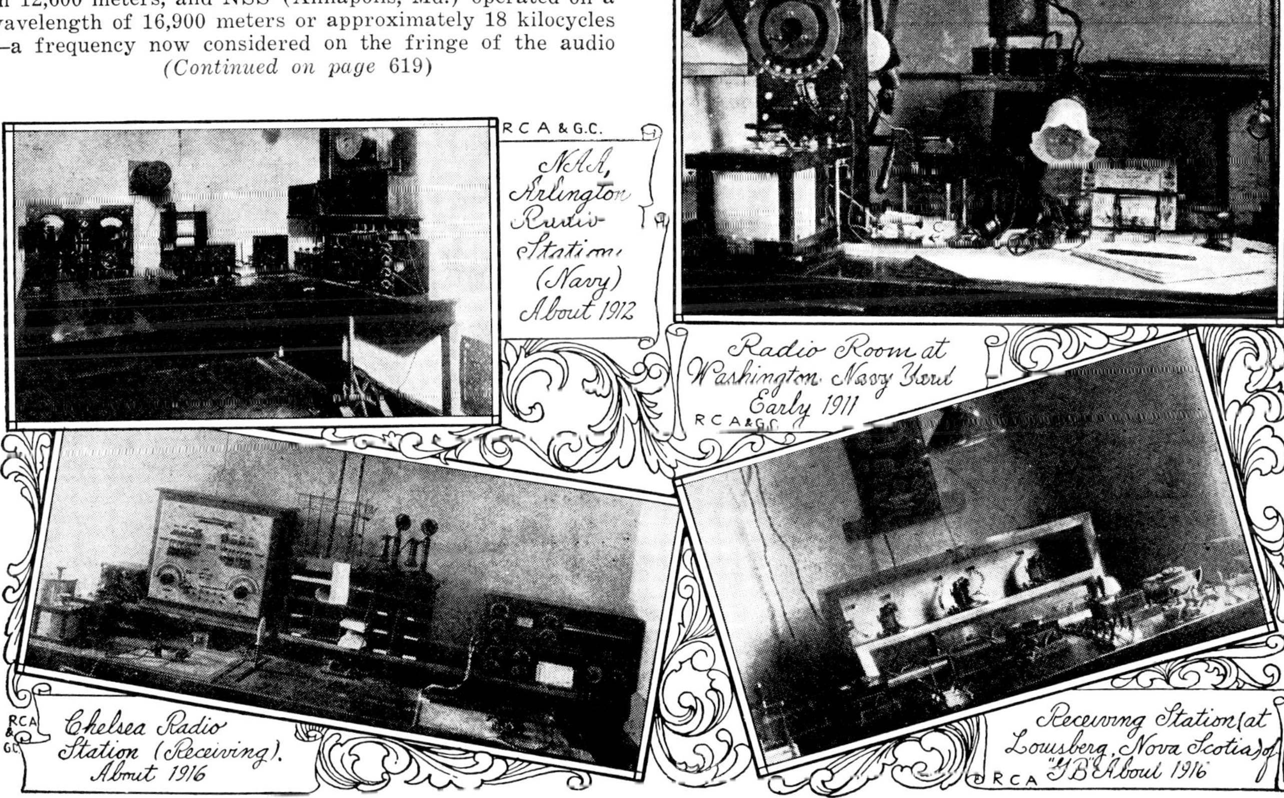

RCA & G.C. NAA, Arlington Radio Station (Navy) About 1912

Radio Room at Washington Navy Yard Early 1911 RCA & G.C.

RCA & GC Chelsea Radio Station (Receiving). About 1916

Receiving Station (at Lousberg, Nova Scotia) of "YB" About 1916 RCA

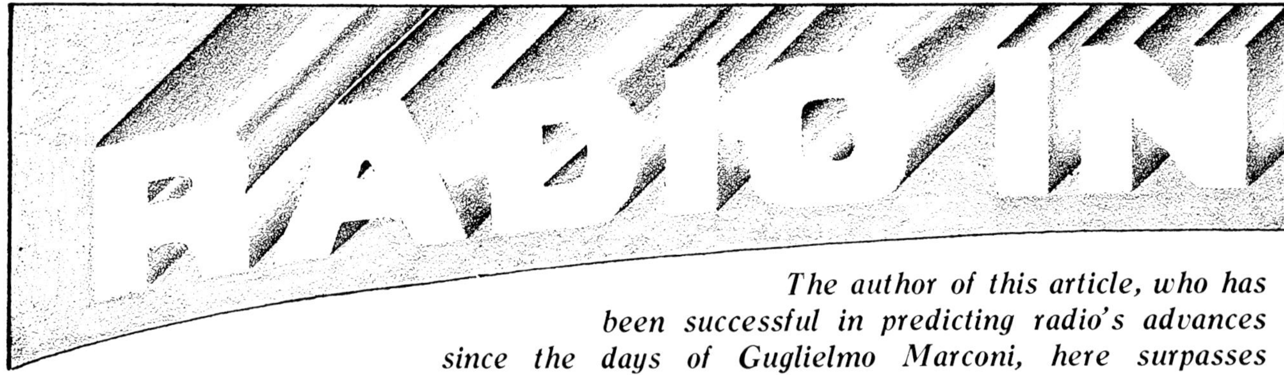

The author of this article, who has been successful in predicting radio's advances since the days of Guglielmo Marconi, here surpasses

HUGO

SO much has been written about the supposed future of radio that forecasting often becomes a commonplace gesture, and frequently the predictions are no longer taken seriously either by the radio industry or by the public.

I have never taken any particular pride in all the many predictions which I have made in the past (many of which have come true) because, after all, others could have done the same thing if they had thought long enough about the subject, and logically plotted a curve of the future. Such predictions are not as difficult as it might appear, nor are they particularly miraculous either. Any industry, and science in particular, always follows a well-planned path which can be predicted

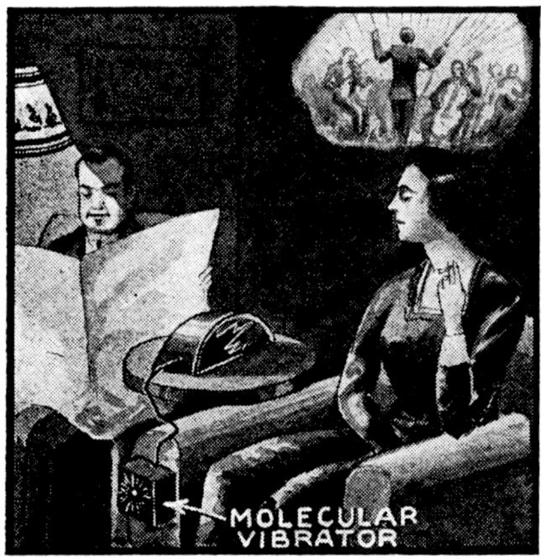

No sound will be heard an inch from the receiver!

with fair accuracy by almost anyone who takes the trouble to do so.

Radio is no different, and what prognostications I have made in the past were always based upon actual needs of the industry; that is the only reason why I have lived to see the fulfillment of many of my predictions.

What I have to say in the following paragraphs follows along the same plan, and while some of the ideas may sound impractical at the present time, remember that many of the other predictions which I made sounded equally foolish and preposterous when I made them.

As I have stated before on many occasions, radio is still in its earliest infancy. As a matter of fact, the entire art remains extremely crude. Perhaps in 100 years or so, radio will really become a great art, which it certainly is not today.

Many problems remain to be solved because most of the instrumentalities which we use in radio today are exceedingly infantile. In other words, I deliberately make the statement that radio still is in the cradle, and the lusty young infant has not as yet started to crawl, let alone walk. To be sure, we have advanced a great deal in the art, but not more so than a six-months-old infant has progressed along the road toward manhood.

While we have made some progress with interference between stations, and those that overlap, much remains to be done. While we think that today's sets, particularly those of the superheterodyne variety, tune sharply, the next two decades will show a complete revolution in this phase alone. Radio will not be satisfactory until the day when there will be no cross-talk between stations themselves, and even no heterodynes from distant stations operating on almost the same frequency.

It is quite likely that our entire ideas as to tuning will be thrown overboard twenty years hence, and that we will use entirely different means to accomplish tuning. The day may come when we will use neither the present type of tuning inductance nor today's type of condenser. Indeed, we may not use inductances or condensers at all in the present accepted sense of the word. While we have made headway in the elimination of man-made, as well as natural, static, we have done so at the expense of power. In other words, we have increased the watts output of our radio stations in the firm belief that by having great enough power, the emitted signal from the station will be sufficiently strong to override and blot out whatever static there is. This is akin to shouting at the top of our voice in order to make a partly deaf man hear us, when it might be much better to otherwise devote our energies to improving a man's hearing, and then converse in a normal tone of voice.

In other words, the power of the broadcast stations has lulled our radio engineers and designers into a mental state wherein many think that the problem is really solved when, indeed, experience shows that real engineering work has to be done in the radio set itself by means of filters or other devices. The very fact that our radio sets are becoming more sensitive all the

No matter where you may be, you will be reached—

time makes a crying need for noise elimination not only out in space, but within the set itself, because we have not only noises from static but from within the tubes. All of these problems deserve much more attention than they get today.

Our broadcasters are spending untold fortunes on the very best of programs, and the finest music that was ever heard on this planet. So when we listen to a Toscanini concert it is the height of absurdity to have such a program

Learn while you sleep, in your radio receptor bed!

his former achievements by stating along what lines the ramified art of radio will develop in the not-so-distant future of the industry.

GERNSBACK

marred with all sorts of noises which, believe it or not, will no longer be heard in a radio set 25 years hence. It means that new principles and new discoveries will be made to slay the monster—Noise, and I have the firm belief that the time

—through micro-wave broadcasting and pocket sets!

will come when we will really have noise-free radio receivers.

This brings us to the radio set itself. We are now going through a curious cycle of development. When we had no loudspeakers we bemoaned the fact that we had to sit at our sets with headphone receivers clamped on our ears. Then when the loudspeakers finally freed us from the head gear, bedlam was let loose in an otherwise quiet world, and the thing for which we wished so ardently has now become

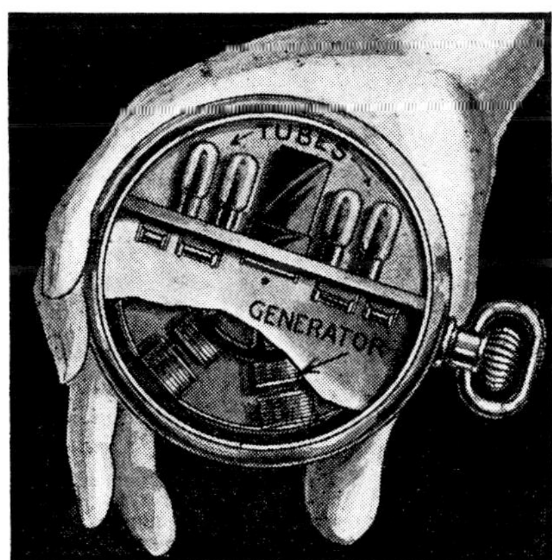

Your watch-size pocket receiver will be self-powered!

an annoying pest which again threatens radio because not everybody wishes to listen to a radio set going at full blast. As civilization advances our nerves are getting more and more jangled and what we really want is a "silent" radio receiver. In other words, we want a radio set which we can enjoy *personally* without arousing the entire household. Then too, we may wish to listen to programs in bed when we are ill, or otherwise without annoying others in the room. While it may be said that some progress has been made along these lines, still it goes right back to the earphone, which does not solve the problem. Thus we may have a telephone receiver in our pillow and we listen to the music in this manner, although this is old stuff because it was done long before the advent of the loudspeaker.

There are, however, other ways of getting the music and sounds to our inner consciousness without taking advantage of loudspeakers and telephone receivers. There is such a thing known as *molecular vibration.* I can see in the future, furniture such as chairs or beds wired with a special apparatus or instrument which will give off sonorous sounds the instant one touches the chair or the bed merely with the fingertips. The idea here is to recreate sound in the bony structure of the human skeleton in such a manner that no one except the person making contact will hear the sounds. Anyone standing even one foot away from him will hear nothing. The method which I have in mind is not an electromagnetic sound reproducer working on the principle of the *osophone* which I patented years ago, which again is nothing but a special telephone receiver. Rather the method which I have in mind is purely *molecular.* Years ago I witnessed some experiments where I saw an electric motor speak and talk by molecular vibration. There were no diaphragms, no vibrating armatures of any kind, but the sound was produced by means of molecular vibration.

When radio manufacturers adapt this principle, we will be able to enjoy radio as we have never been able to enjoy it before, and at the same time we will not annoy others when they wish quiet.

Many years ago I imagined an instrument which I termed the Hypnobioscope. This fanciful instrument was supposed to teach the student while he slept. Much to my surprise this idea is already being utilized to teach students the code when asleep, even though they could not learn it when awake. (See article entitled "Predictions and Fulfillments" in this issue.) The time is not far away when this will actually come into universal use, and again by means of the above-mentioned molecular radio sound transmitter. By that time broadcast stations will give special programs at night, and stations will vie with one another to give important instruction, whether it be languages, mathematics or history, during the hours from 12 M. to 5 in the morning.

We will not need to wear head receivers or, in fact, anything at all, yet the radio program "connected" to our beds will be transmitted to our inner consciousness with sufficient volume—but not too much volume—to avoid awakening us. As I pointed out before in connection with the Hypnobioscope, the human mind is unusually recep-

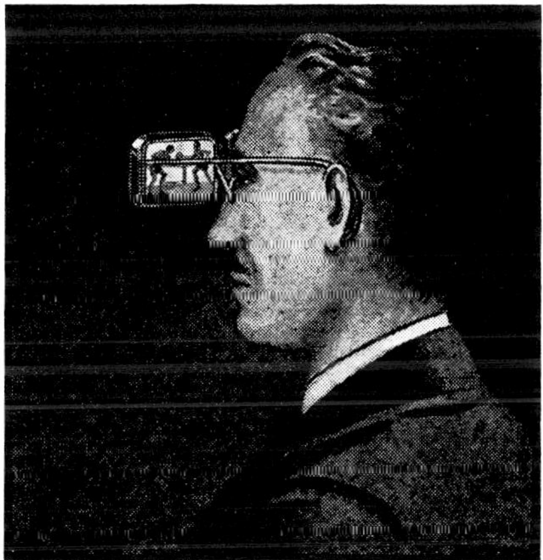

Television spectacles will show 3-dimension images!

tive while it rests, and while one sleeps. Indeed, the "learn while you sleep" method proved itself at the Pensacola Naval Training Station, because students who never could learn the code while awake, found it simple while asleep.

Thus, in the future, you will find that our knowledge will be tremendously increased by the "learn while you sleep" method, via radio. Indeed, it would not surprise me in the least if

(*Continued on page* 613)

PREDICTIONS AND

An editor who takes his occupation seriously is in position to dispense to his readers largesse in the form of prophecies upon which foresighted persons may capitalize. By observing world-wide technical trends and noting public reaction to

ROBERT EICHBERG

N his writings over a period of more than 30 years, Hugo Gernsback, editor of *Radio-Craft*, has foretold a number of important trends in various scientific fields. Not all of his predictions, of course, have come to pass, but an astonishingly large number have—particularly in the radio industry. In fact, one may examine almost any phase of broadcasting as we now know it, and trace it back to some prognostication made by Hugo Gernsback in the days before "wireless" had become known as radio.

Perhaps the earliest editorial record of an important Gernsback prediction appeared in the February, 1912, edition of his magazine, *Modern Electrics*, which he published and edited at that time. In it he foretold, with incredible accuracy, the provisions of the First Radio Act, a law which was not passed by the Senate until May 7 of the same year. In fact, the bill included all the major recommendations made by Mr. Gernsback in the previously-mentioned article. That is either a remarkable coincidence, an uncanny ability to foretell the future, or Mr. Gernsback virtually dictated the bill. Compare his words with those of the Wireless Act:—

Mr. Gernsback suggested:

"There should be a bill passed restraining the amateur from using too much power, say anything above 1 K.W.

"The wave length of the amateur wireless station should also be regulated in order that only wave lengths from a few meters up to 200 could be used. Wave lengths from 200 to 1,000 meters, the amateurs should not be allowed to use, but they could use any wave length above 1,000 . . . ."

Section 15 of the Wireless Act of 1912 states:

"No private or commercial station not engaged in the transaction of *bona fide* commercial business by radio communication . . . . shall use a transmitting wave length exceeding two hundred meters or a transformer input exceeding one kilowatt except by special authority of the Secretary of Commerce and Labor contained in the license of the station . . . ."

Save for the use of the waves in excess of 1,000 meters, the bill included all the provisions prescribed by Gernsback.

But while that gentleman looks upon this as his outstanding achievement in the field of radio prognostication, it is by no means his earliest nor, to the minds of some others, his most astonishing.

His predictions, indeed, may be traced back to as early a date as May, 1909, when, in *Modern Electrics*, he foretold the possibility of chain station operation—and that was in the days before broadcasting existed as such, let alone chains.

Nor, for that matter, were the prognostications of Hugo Gernsback confined to the austere editorial. In a serial science-fiction story, which commenced in the April, 1911, issue of *Modern Electrics*, he narrated the adventures of a character—"Ralph 124C41+"—who was imagined to be living in the year 2660 A.D. The instalment in the June issue of that year carried the following:

"He soon went to bed, but before he fell asleep he attached to his head a double leather head-band with metal temple plates . . . . He then called for his faithful butler and told him to 'put on' Homer's *Odyssey* for the night . . . . The butler . . . . took down . . . . a narrow box . . . . From

THE MODERN PUBLISHING CO.
H. GERNSBACK - - - President and Secretary
O. J. RIDENOUR - - Vice President and Treasurer

Entered as second class matter March 31, 1908, at the New York Post Office, under the Act of Congress of March 3, 1879.

Vol. IV. FEBRUARY No. 11

EDITORIAL.

There should be a bill passed restraining the amateur from using too much power, say, anything above 1 K.W.

The wave length of the amateur wireless station should also be regulated in order that only wave lengths from a few metres up to 200 could be used. Wave lengths of from 200 to 1,000 metres, the amateurs should not be allowed to use, but they could use any wave length above 1,000. If this is done, all interference with Government, as well as commercial station, will be done away with and the wireless situation will then be the same as to-day. The amateurs will have the same liberty and perhaps greater liberty than to-day, and complaints against them from Government or Commercial stations will cease automatically.

(1)

(1A)

THE MODERN PUBLISHING CO.
H. GERNSBACK - - - President and Secretary
O. J. RIDENOUR - - Vice President and Treasurer

Entered as second class matter March 31, 1908, at the New York Post Office, under the Act of Congress of March 3, 1879.

Vol. V. JUNE No. 3

AS predicted in our last issue the Alexander (and Bourne) bill passed the Senate.

The new bill (S. 6412) passed the Senate on May 7th, and is now fairly well under way to becoming law.

Turning to Paragraph No. 15—the one of importance to the amateur,—we find that he must not use a greater wave length than 200 meters, nor use a greater power than a 1 K.W., input in either his transformer or spark coil,—nor more than ½ K.W. if he is within five nautical miles of a government station.

400,000 WIRELESS AMATEURS.
To Discourage Their Work Would Check Progress in the Art.

To the Editor of The New York Times:

I note your timely article on "The Wireless Control Bill." While you have taken your stand very well, as far as the commercial interests go, I would like to give the views of the wireless experimenters and amateurs whom I have represented for the last six years.

Much has been written and said against the wireless amateur, and while it is true that some mischief has been done, report has never been made of any case where serious damage was done, except in a single instance a few years ago, when a Massachusetts amateur transmitted a false "S. O. S." call, which sent a United States Government boat out on a fool's errand. On the other hand, the amateur has done a great deal of good by taking up distress calls, and it should be understood that good work has frequently been done in the past by amateurs receiving and transmitting such calls where they would probably not have been received by the few commercial and Government stations. Furthermore, a great many amateurs, while not necessarily schoolboys, have done much to further the art in general, and many patents have been taken out during the last three years by such students interested in wireless.

The interests which would like to see the Alexander bill passed have evidently not looked very much into the future. On account of the vast distances in this country, radio-telegraphic and radio-telephone intercommunication will positively find a distinct usefulness, greater without doubt than that of the present telephone. An immense usefulness alone will be found in radio-telephone stations between moving vehicles and fixed posts.

H. GERNSBACK.
Business Manager Wireless Association of America, Editor Modern Electrics.
New York, March 27, 1912.

(2)

(3)

(4)

(1) The famous Gernsback editorial of February, 1912, suggesting the provisions which subsequently were to become the first Radio Act. (1a) Published 4 months later, this tells of the passage of the Act, virtually as outlined in the editorial. (2) A plea for the rights of the amateurs by Mr. Gernsback, published in the *N. Y. Times* of March 27, 1912. (3) "Learn while you sleep" method predicted in cover painting illustrating a Gernsback article featured in December, 1921, *Science and Invention*. (4) The idea in use at the U.S. Naval Training station at Pensacola, Florida, in 1922, to teach code to student operators.

FULFILLMENTS

existing facilities he is able to offer tangible and workable suggestions by which everyone may profit. Hugo Gernsback, whose liberality in suggesting worthwhile improvements in the radio field is unequalled, is one such editor.

this he extracted a large but thin reel on which was wound a long, narrow film. This film was entirely black but for a white transparent wavy line running in the center of it . . . . He introduced the end of the film into the *Hypnobioscope* . . . . which transmits words direct to the sleeping brain in such a manner that everything can be remembered in detail the next morning."

So wrote Mr. Gernsback in his fictional story. He repeated this prediction, somewhat more seriously, in an article which was featured in the December, 1921, issue of *Science and Invention*.

Then, in October, 1923, *Radio News* carried the startling information that Gernsback's prediction had come true, for it printed an article by J. N. Phinney, Chief Radioman, U.S.N., in which the practical application of the idea was described. Mr. Phinney had acquired rapidity in code by sleeping next to a telegraph sounder. In 1922, he had conducted tests, sending code to 12 sleeping students who wished to learn radio. The students wore, while they slept, aviation-type helmets containing headphones. The tests were conducted for some period and all reported improvement. The "*Hypnobioscope*", which Mr. Gernsback had conceived, had become an actuality—without waiting for the year 2660 either!

In the *Electrical Experimenter* of December, 1918, he predicted that wireless signals would be heard around the world, and in July, 1919, inaugurating his *Radio Amateur News*, Mr. Gernsback asserted his faith in the future of radio in America; then in the last issue of that publication, before it became *Radio News*, he brought up three extremely interesting points, for in the May, 1920, issue, he wrote editorially: (1) "There is a tremendous market existing for a small (send-receive outfit that can be put into a suitcase and which works on 6 dry-cells. Why does not someone invent it?"; (2) "Then there are our good old telephone receivers. Year after year, we have been using them, whereas it has often been pointed out that we really do not want them because they tire our heads and ears, and are unsightly and very inefficient."; (3) "Why must we receive by ear? Why not, for instance, by sight? It can be done, and at least as efficiently as by ear." The proof that these predictions were well founded is borne out by (1) the portable transceivers of today, (2) the loudspeaker, and (3) the oscilloscope or "Tuning Eye" indicator.

In June of 1920, this magazine became *Radio News*, the publication from which the following predictions are taken, unless otherwise noted.

The July, 1920, issue contained more than one such suggestion. For example, Mr. Gernsback wrote, "Twenty-five years from now our long-distance stations will be operating with comparatively small power . . . . It should be possible in 1945 to telegraph 12,000 miles . . . . with a power which does not exceed ½ k.w." This is borne out in the results achieved by numerous amateur short-wave stations. He also wrote, "And one of these days, we will wake up and find that some genius has made it possible for us to see actual radio waves." The present-day cathode-ray oscilloscope has done that. He also said, "One of the coming wonders without doubt is radio movies." Although Mr. Gernsback had spoken of television many years earlier (indeed, he is credited with having coined the word in 1909) when he said it would be possible to transmit images "without using numerous wires", this seems to be the first

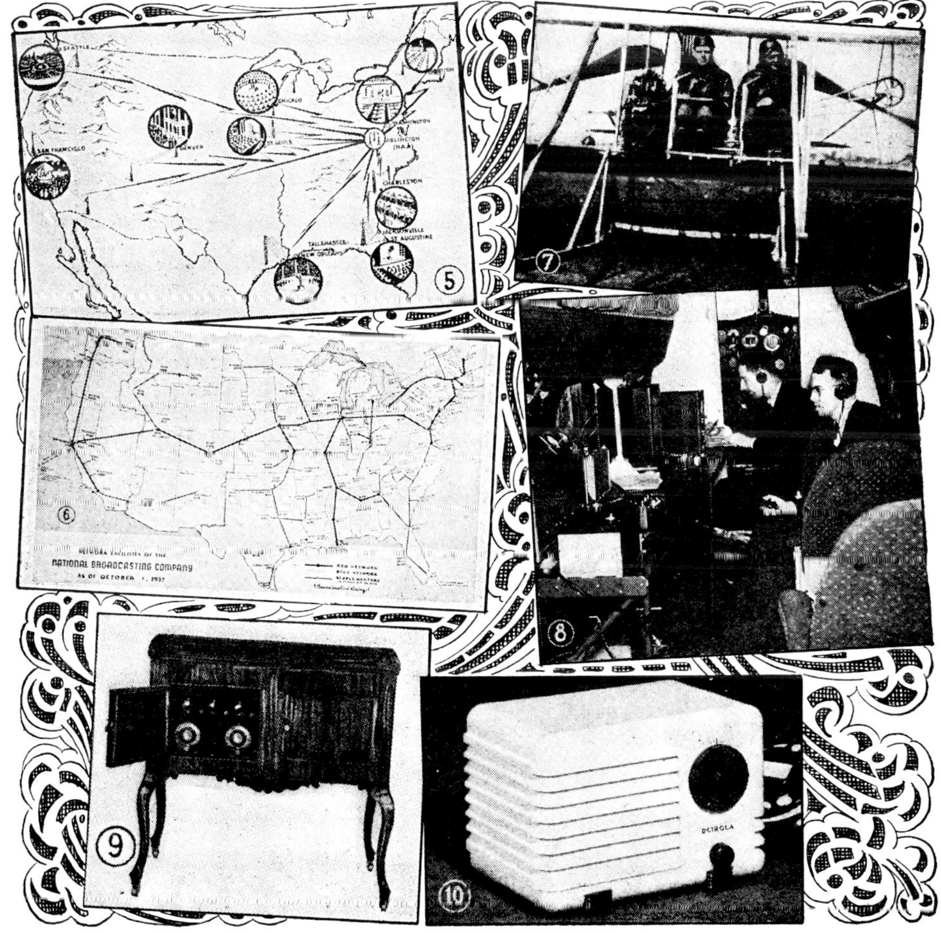

(5) Station networks, as predicted in *Modern Electrics* for May, 1909, and sketched in September, 1916. (6) An example of the modern network, as shown on a N.B.C. map for October, 1937. (7) A U.S. Army photo of a 1912 plane, with radiotelegraphy, as predicted. (8) A United Air Lines plane, with an elaborate phone installation, as Mr. Gernsback foretold. (9) In January, 1922, he predicted the development of sets housed in console cabinets; this one was advertised 7 months later. (10) Previously he had called for smaller models; this 1938 midget is the answer to his prediction of 1921.

suggestion to be made of televised movies.

In the September, 1920, issue, after having discussed radio concerts, Gernsback went on to say, "Why cannot someone go after the Presidential candidates and invite them to make a speech via radio thru a powerful telephone apparatus in the near future. With proper advertising and with the proper enterprise behind such a scheme, it certainly should not cost a great deal to do. The people of the United States . . . . would get a chance to listen to our candidates in a very novel manner." This was a follow-through on an earlier idea, which Gernsback had expressed in *Modern Electrics* as early as September, 1916, when he suggested that it would be a fine thing if President Wilson would use the radio to address all the United States simultaneously. The collator wonders if Mr. Gernsback, around election time, ever regrets his suggestion of 1920.

When midget sets and loudspeakers were virtually unknown, in February of 1921, Mr. Gernsback predicted that by 1931 there would be compact radio receivers which would include one or two stages of audio and said that ". . . . the sounds will come right from a horn similar to our phonograph today, only built along miniature lines . . . . The small outfit is *the* thing and we cannot urge manufacturers (too strongly) to pay attention to this phase."

The following month he foretold the end of spark transmission for amateur operation, saying, "Have you ever stopped to think how clear the nightly ether would be if all the amateurs were using radiofone equipment instead of wireless telegraphy . . . . The writer believes that the radiofone, as far as the amateur is concerned, is the right way out." While amateurs still use I.C.W. as well as C.W., they have discontinued spark transmission, which was what Mr. Gernsback had reference to.

In May he clarified the point somewhat in an article which concluded: "Sooner or later we will be forced to give up our spark stations and operate on C.W., and this is as it should be. The sooner we amateurs realise this, the better it will be for us . . . ."

The condenser-type loudspeaker, which flourished for a brief while, was prognosticated by Gernsback in October, 1921, and the following month the transmission of newspapers by facsimile was suggested. Eight years later the present collator actually did this for the *New York Evening World*, sending "the printed page, type, pictures and all, in less than a half-hour," just as Mr. Gernsback had predicted would be done.

One prediction, made in an editorial written for December 1921, seems particularly important. Mr. Gernsback said, "We can not expect the future wireless radio enthusiast to buy the parts of an outfit and put it up himself . . . . What he wants is a sort of '*radiotrola*' that is all connected up for him and, with a few intelligible instructions, can be used as easily by the young hopeful as by granddad . . . . All we require is a nice cabinet which contains all the instruments readily put up. In front there should be a knob that could be turned for any desired entertainment, such as jazz, lectures, good-night stories for children, grand opera, etc. . . . ." Remember, we had no complete radio sets or consoles in 1921.

Mr. Gernsback opened 1922 with a prediction in the January issue of *Radio News*, saying, "In 1922, we may safely predict that millions of homes will have their radio receiving sets right in their parlors . . . . The '*radiotrola*' of which we spoke last month will come into its own. The latter for the time being will eclipse, or put into the shade, its predecessor, the Victrola and the

(*Continued on page* 610)

(11) A 1912 Pierce-Arrow limousine, equipped with electro-telephonic communication from tonneau to chauffeur's seat—an early forerunner of auto-radio. (12) Instrument panel of modern Ford automobile, showing dial and controls of present-day auto-radio receiver as installed in this make of car. (13) An old Bristol loudspeaker, of a type then considered one of the best possible to produce. Gernsback wrote that larger diaphragms were needed, especially for faithful reproduction of bass notes; a short time later, manufacturers were marketing speakers with 3-foot diaphragms, (14) an Ensco wall-model is shown. (15) The Jenkins televisor of 1931 used a Nipkow-type pinhole disc to scan the plate of a neon tube, the image being viewed through a magnifying lens. Gernsback suggested that this be replaced by a Braun (cathode-ray) tube. The 1938 Philco receiver (16) uses the tube he had described, as do most other present television sets which are designed for home use. (17) Modern short-wave radio therapy, as foretold by Gernsback in a 1924 editorial.

NEW CIRCUITS IN MODERN RADIO RECEIVERS

The details of the modern radio receiver circuits that make them "different" from previous designs are illustrated and described each month by a well-known technician.

F. L. SPRAYBERRY No. 6

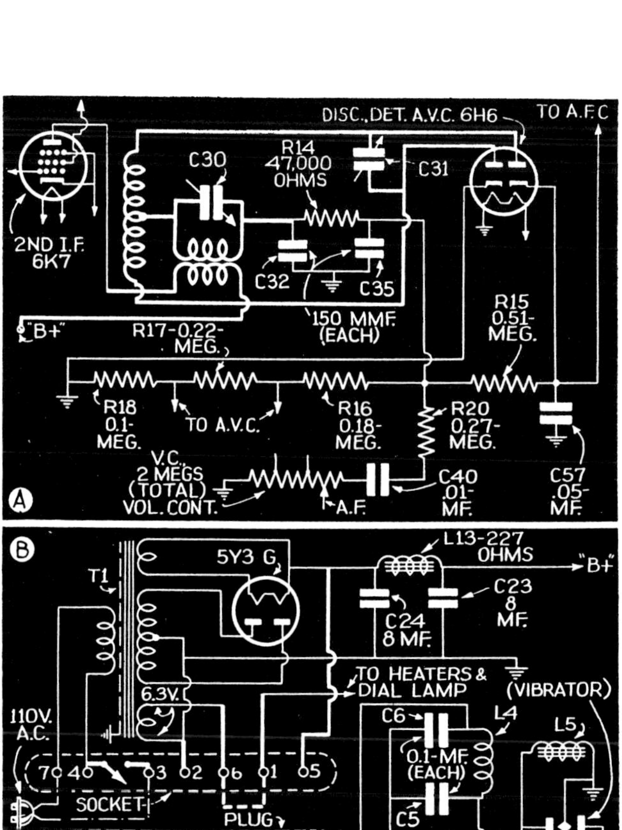

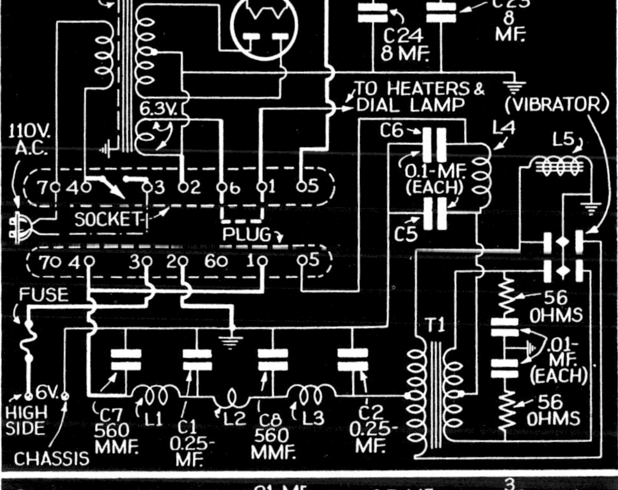

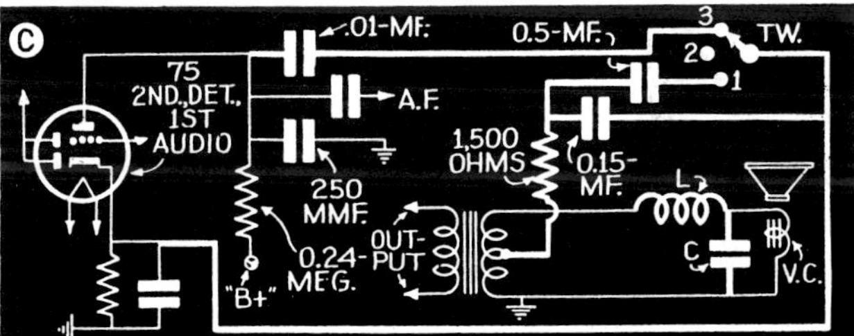

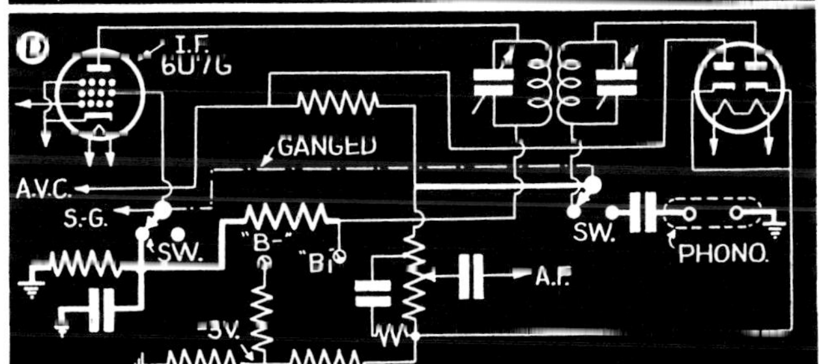

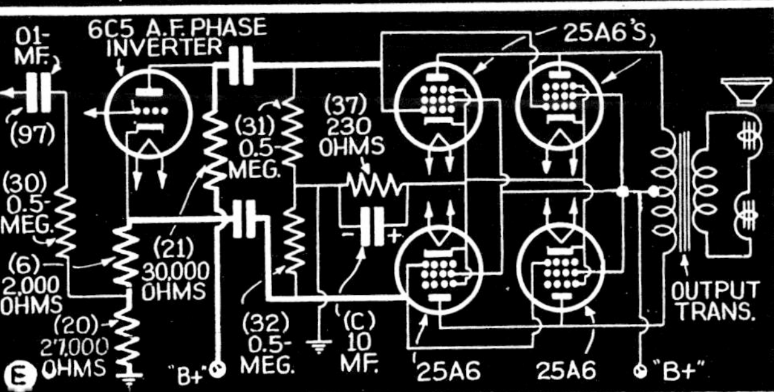

Fig. 1. The heavy lines in the circuits are the points discussed in the text.

(1) NEW INPUT DISCRIMINATOR CIRCUIT

General Electric Model F-135. An I.F. signal voltage is developed across the circuit tuned by C30, Fig. 1A, from the plate coil of the last I.F. amplifier. This voltage is applied to both diode plates of the discriminator detector "in-phase", from the electrical center of the coil each end of which attaches to one diode plate. With the I.F. trimmer C31, an I.F. tuned circuit is formed between the plates. The diode loads, however, do not have similar impedance values as one (the A.V.C. and signal load) is practically a pure resistance while the other (the A.F.C. load) is largely capacity reactance due to C57. This latter impedance being lower than the former one, more current tends to flow in the upper-half of the coil than in the lower-half. Thus, the upper diode plate voltage changes more with the signal than the lower one. This difference in diode plate potential gives rise to a series resonant current in the diode input tuned circuit, which results in a voltage of resonant frequency being fed to the plates out-of-phase. The shift in phase, of the "in-phase" signal voltage to that of the "out-of-phase" voltages as the signal frequency varies, creates the A.F.C. voltage in the customary fashion.

(2) SET USES A.C. OR BATTERY SUPPLY

RCA Model 85BT6. Use this set as a 110-volt, A.C.-operated home receiver or simply plug it into a vibrator power supply operating from a 6-volt storage battery. From the conventional A.C. power unit are brought leads as in Fig. 1B, to a 7-hole socket mounted on the left-rear of the chassis. The vibrator power supply unit is separate and is provided with a cord and plug which plugs into the socket for battery operation. Its plug pins are identified in Fig. 1B by numbers corresponding to those in the receiver socket mounted on the receiver chassis. It will be observed that the same controls are used for either power supply.

(3) FEEDBACK-TYPE TONE CONTROL

Philco Model S-1526 (Automotive). Control of the amplitude characteristics of the A.F. amplifier within the audio spectrum for voice and music, is accomplished without the sacrifice of an overall signal reduction by the means shown in Fig. 1C. In position 1 of the tone control switch TW, the audio response is essentially "flat" and extended appreciably at each end (low and high frequencies). It is actually rising slightly at either end. In position 2, the low frequencies are attenuated to a satisfactory extent and in position 3, the high frequencies in addition are attenuated. The average center part of the spectrum (around 400 cycles), however, remains at the same level throughout. A low-pass filter L-C is used in the speaker voice coil circuit so that extremely high frequencies beyond the control range will not be reproduced.

(4) NEW PHONOGRAPH CHANGEOVER

Stewart-Warner Models 1821 to 1829. To insure that the radio signal is completely eliminated when set for phonograph reproduction this circuit (see Fig. 1D) disconnects the 1st-detector and I.F. screen-grids, and the signal-diode plate return. In this way, the input section of the receiver

(Continued on page 609)

THE RADIO MONTH

(Photo—Westinghouse)
Drs. Marden and Beese make photoelectric test of high-pressure mercury arc, brighter than the sun (1,100,000 candle-power to square inch). Observe water-cooling tubes. It is expected to give large-screen television a boost.

(Photo—W. E. Shrage)
Movable speaker for Berlin theatre capable of handling 1,000 watts is largest hi-fi unit in world. It weighs a ton and reproduces faithfully from 10,000 down to 40 cycles.

RADIOFFICIAL REGULATIONS

MEETING in Nice (France) last month (it's a nice place to meet in) the Union Internationale de Radiodiffusion, which deals with European broadcasting, voiced the difficulties of continual interference on that crowded continent of none too amicable independent nationalities. As to unwanted interference, directional antennas are helping to reduce that. It was recommended that a program of exchanging broadcasts of chimes for New Year's, 1938-'9, be extended to further interchange of records. And it was voted unanimously that the European public would not be satisfied with the idea of advertising-sponsored programs, instead of making payment for their programs in the form of set licenses. (Two American broadcast networks were represented at the meeting.)

The Federal Trade Commission, laying down regulations against misrepresentation in radio sales, allows a 540-18,000 kc. set to be advertised and sold as "Limited All-Wave." It also clamped down last month on a "radio metal locator," and required the advertiser to cease representing that it could "distinguish between metals such as gold and silver."

The F.C.C. issued an order requiring owners of airplane radio equipment to have station licenses, and to have the frequency checked periodically at authorized laboratories. Further, after a long hearing, it last month suspended from practice an attorney for filing before it the applications of "dummy corporations" seeking station licenses. A similar position was taken by the Federal Court of Appeals, D.C., disapproving incorporation to hide ownership of licensees.

At a meeting in Chicago of those interested in "educational broadcasting" last month, F.C.Commissioner Payne observed that "Program standards must be raised," and that this would be impossible if left to broadcasters, appealing to the masses. Rebuttal by Merrill Denison said that radio cannot "inflict again a high school and college course on listeners." (Many groans are heard from British listeners about educational programs of B.B.C.—and none from listeners in other European states where they take what the government thinks is good for them, and like it.) Controversy will undoubtedly continue far into the century.

Regulations reported last month by U.S. foreign trade bureau include: Cuba prohibits radio sets in busses, public or private; Turkey requires purchaser of set to be given "prospectus" and diagram, copy of which is filed with government. Sydney, Australia, lighting service equips 42 service vehicles with 2-way telephone service. Each carries also emergency signal device on transmitter to acknowledge message, or request repetition, when voice cannot be heard at central or receiving relay stations.

The competition of foreign language broadcasts by different powers seeking to spread propaganda to colonies and partisans in other nations was admitted by the British Empire when the B.B.C. announced, last month, it will broadcast in Spanish and Portuguese to South America and in Arabic to the Near East. "Not in German and Italian as yet." And in this country, two stations in a large city, which had been keeping on different languages at the same hour, severed diplomatic relations and concentrated on Yiddish from 8 to 9 p.m.

Japanese government installed amplifiers on electric trains, with loudspeakers to inform passengers of station stops, etc. Novel feature of an-

(Photo—Rock Island Lines)
New streamlined "Rocket" trains, covering the mid-West at 70 m.p.h., have radio built into their light stainless-steel coaches, diners and parlor cars. The set (upper-right insert) is in diner. One speaker is shown over coach door.

(Photo—Bell Telephone Laboratories)
At Point Reyes, Calif., this station keeps continuous watch for trans-Pacific messages on 22 channels. The radio receivers are alternately switched, automatically, to fixed H.-F. amplifiers.

IN REVIEW

Radio is now such a vast and diversified art it becomes necessary to make a general survey of important monthly developments. RADIO-CRAFT analyzes these developments and here presents them in review.

nouncement last month is that "it will be especially useful in case of accident or air-defense maneuvers" when lights are turned out to avert bombing.

TELEVISION WAITS ON STANDARDS

RUMORS last month of television licensing caused inquiry, and the following definite statement from the office of the F.C.C. was received by *Radio-Craft:* "So long as there are no radiations or violations of the experimental and non-commercial status of co-axial cables and television broadcast stations, the Commission has no jurisdiction. Though future activity cannot be predicted at the present time, it would appear that commercial television will not be practical until some form of standardization is adopted. No applications have been filed for a commercial television authorization; and no action will be forthcoming *(from the Commission)* until there is a satisfactory showing that television is ready to be presented to the public."

Something like the general idea of television—direct sight of impulses—is provided in a patent (No. 2,101,139) issued last month to Clarence W. Hansell and assigned to RCA. It provides a viewing screen covered with copper sulphate, in a moist atmosphere; which turns white when ultra-short radio waves impinge upon it, either by reflection or direct from a transmitter. *Thus, it is proposed, a ship or plane may be seen through fog or darkness!* It does not appear how practical the device is, in its present state of development.

Ship reception of television signals, considered in the light of present medical service by radio from shore to ship, caused proposal last month by *Reynolds* (London) that it may yet be possible to enable a doctor to see a patient at sea, and (even more practicable) to illustrate by action the operation or treatment which he recommends a seaman's companions to administer in an emergency. Even appendicitis operations have been performed on radio advice; but how much better if they could be first illustrated to the amateur surgeons.

Televised news service appeared commercially last month, when Peck Television installed a receiver in Jack Dempsey's (popular N. Y. restaurant) demonstrating mirror-scanning of signals transmitted over leased lines. Letters 4 inches high run across a screen, giving about 60 words a minute, and can be read 75 feet away. Plan is to install receivers where news bulletins, etc., are wanted in public places. Master tape is typed at transmitter

(Continued on page 648)

(Photo—Globe)

Unveiling a granite monument at Poldhu Cove, Cornwall, England, to commemorate sending of first radio message across the Atlantic on Dec. 12, 1901, by Marconi and associates.

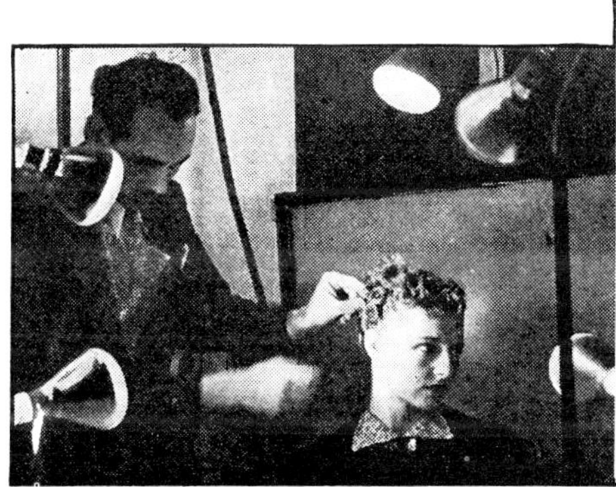

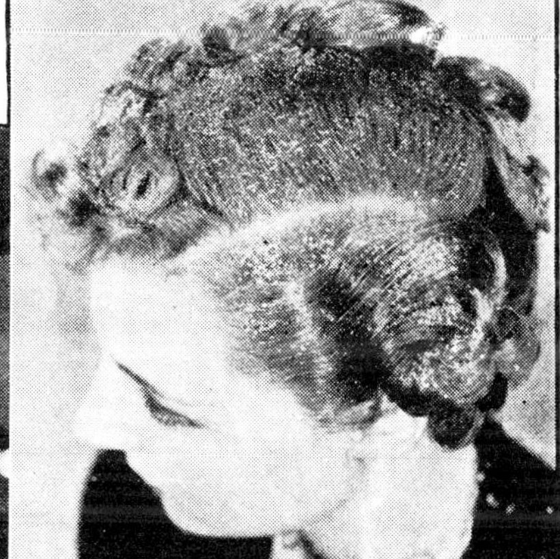

(Photos—NBC, by William Haussler)

Last month's all-fashion demonstration of television at studios in Radio City, New York, was made before stylists, designers, fashion writers, and store executives. It was a test, not only of television, but of televisibility of various additions to costume and coiffure. Here we have an expert applying glittering powder ("star-dust") and rosettes to the "mannequin's" hair.

(Photo—Globe)

Television set in the Cecil (Strand), high-class London "pub" (bar and grill), entertains customers in mid-evening. Problem of entertainment tax will probably next be raised by the Internal Revenue Department (they would!).

Right, radio equipment of "airplane hotel" ship Friesenland, stationed between the Azores and South America.

Any of 6 all-W. and S.-W. receivers can be operated from any desk, covering 15 to 20,000 meters.

(Photos—Shrage)

Above, the Friesenland's catapult has just thrown a plane into the air; note crane used to hoist planes out of the sea. The ship is equipped also with course-guiding radio transmitters.

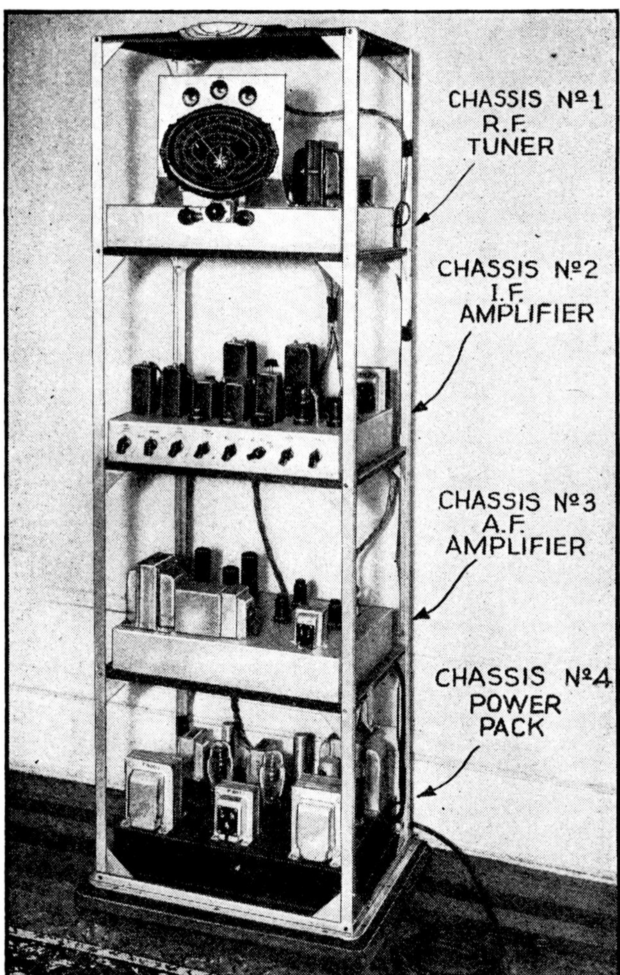

Fig. K. Front view of all chassis of completed 30-Tube Set.

HOW TO MAKE THE RADIO-CRAFT SUPER-DELUXE 30-TUBE SET

Here, at last, is the concluding chapter on this superlative receiver, dealing with the loudspeaker system and the construction of the framework of the complete 4-chassis assembly. *Radio-Craft* will be pleased to receive reports from readers who built this receiver in easy stages as described.

PART VI

IN THIS, the concluding instalment of the series of articles, we will describe the construction of the *rack frame* which houses the 4 chassis. At the time of going to press, the special infinite-baffle cabinet was not completed. Inasmuch as the speaker, or speakers used, will vary with the requirements of the builder, we give in Fig. 17 the dimensions of the loudspeaker cabinet (which is separate from the receiver cabinet) which we believe to be the best. The dimensions can be scaled down to suit the size which the builder requires.

In any event, the cabinet will be better, acoustically, than a flat baffle of equivalent front area.

The rack frame was built open all around (see Figs. K and L) in order to show the construction and placement of the shelves. Veneer panels may be affixed to the front and sides, if an enclosed type of rack is desired.

The first job is to prepare the base platform. See Fig. 18A in which the dimensions are given. It is desirable to use a solid board at least 1 inch thick. *Before* chamfering the edges it is advisable to locate the area which the aluminum rectangle will occupy on the baseboard. The aluminum rectangle must be assembled according to Fig. 18B and screwed to the base platform. The upright angle girders should next be prepared and mounted, one in each corner of the rectangle.

Within the area of the rectangle is mounted the spacer board. The dimensions are given in Fig. 18C. The reason for using this board is so that the level of the aluminum rectangle is even with the spacer, so that the Heavy-Duty Power Supply Chassis may be slid in and out easily, without having to lift this heavy chassis over the edge of the rectangle. After the spacer board has been screwed down, the shelves are ready for mounting.

The 4 shelves are all alike, the dimensions are given in Fig. 18D. The spacing between shelves is 15 inches between base rectangle and 1st shelf; between 1st and 2nd shelf, 15 inches; between 3rd shelf and top shelf, 14½ inches. Holes should be drilled and countersunk in the angle uprights to accommodate the small woodscrews which hold the shelves to the uprights.

The ornamental strips shown in Fig. 18H are screwed to the top shelf. They may be spaced close or far apart as the builder desires. The shelf brackets, shown in Fig. 18G, are mounted next, by means of 2 small wood screws and one self-tapping screw, which should be placed on the side of the upright for each bracket. The platform and upright brackets are next prepared according to Fig. 18I.

These brackets are mounted on the base rectangle and to the uprights with self-tapping screws. This completes the

SPECIAL NOTICE

AS mentioned in Part I of this series *Radio-Craft* has made arrangements to have tested and repaired *without charge* any *Radio-Craft* Super-Deluxe 30-Tube Radio Receiver, built in accordance with the instructions given in these articles and *using only the specified components*, which fails to function correctly (The constructor though must pay shipping charges both ways.). However, custom-set builders who feel qualified to make substitutions and who then encounter trouble, may also have their sets checked in the event of faulty operation; the cost to the set builder in this instance will be the cost of shipping plus a nominal charge for making the necessary tests and repairs and for such replacement parts as may be necessary to effect perfect results. *You can't lose!*

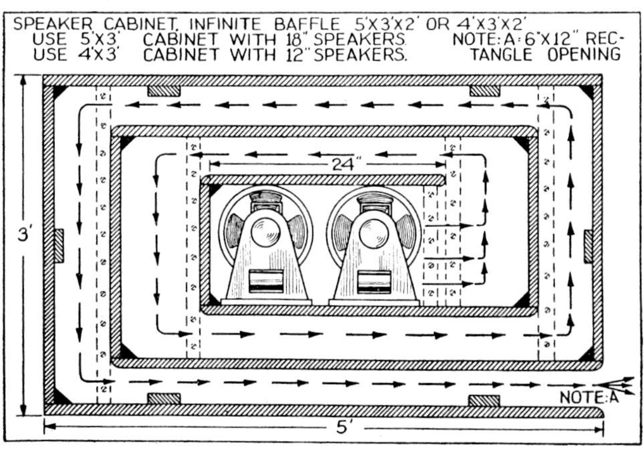

Fig. 17. Constructional data for the infinite-baffle cabinet housing the 2 speakers.

main construction of the rack frame. The A.C. line switch should be mounted on the special bracket shown in Fig. 18F; this bracket is then screwed to the edge of the 1st shelf, on the right-hand side of the framework. The final touch is to prepare and mount the aluminum ornament, shown in Fig. 18J. As can be seen, it consists of 4 pieces of angle aluminum screwed together and to the center of the top shelf.

THE LOUDSPEAKER SYSTEM

The dimensions of the baffle cabinet are given in Fig. 17. The front and rear panels are each one piece of plywood. The rear panel may be ¼-inch thick, while the front panel *must* be at least ¾-inch thick.

The top, bottom and two side panels must also be ¾-inch thick. The sound track within the cabinet may be of celotex or plain pine boards. If a wooden sound track is used each board should be covered with some sound-absorbing material such as felt, hair-wool, kapok or cotton. If celotex is used, the various pieces should all be at least 1 inch thick to obtain overall uniformity.

Two different types of external field supply are shown, which will take care of practically any combination of speakers which the builder may desire to use, in the event that permanent-magnet types of dynamic loudspeakers are considered too expensive. A small field supply which is inexpensive and yet very good is shown in Fig. 19A. Here a 25Z5 rectifier is used and is capable of supplying 100 volts at 100 ma., or 10 watts of field power. Two of these small supplies may be used in parallel, thus operating two fields, each of 1,000 ohms.

If it is desired to actuate a 2,500-ohm field from the same supply, only 40 ma. can flow, producing only 4 watts of field power. This is not really sufficient, so the other type of supply should be used. This is the high-voltage transformer type shown in Fig. 19B and is capable of supplying 300 volts at 200 ma., or a total of 60 watts, which is enough for 3 or 4 large speakers. The speakers which are being used to try out the laboratory model of the 30-tube receiver are two Utah (The Utah Radio Products Co.) 12-inch speakers, model H-12-25 (see Fig. 19C) with a field resistance (each) of 2,500 ohms.

The 9-ohm voice coils are paralleled and hooked into the 4-ohm tap on Chassis No. 3. The field coils are also paralleled, giving a total resistance of 1,300 ohms, and they are connected across a field supply of 200 volts at 150 ma. giving a field strength of 30 watts. This gives each field 15 watts.

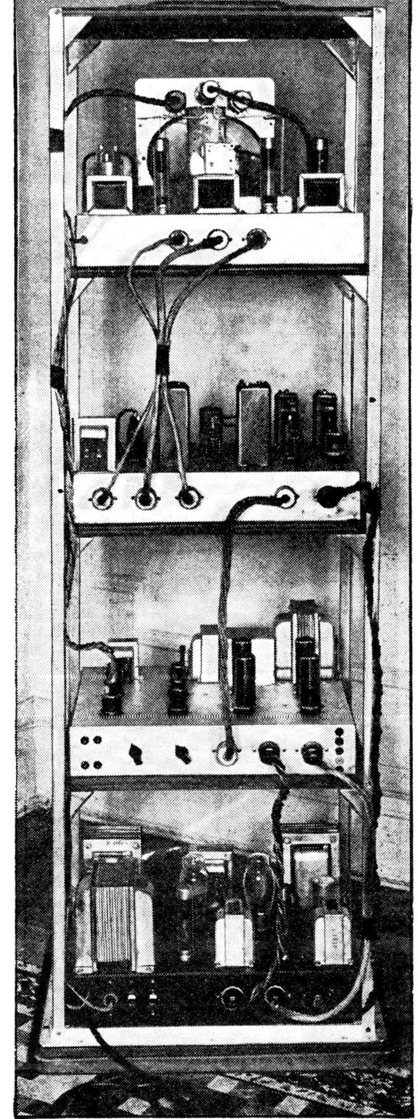

Fig. L. Rear view. Note interconnecting cables.

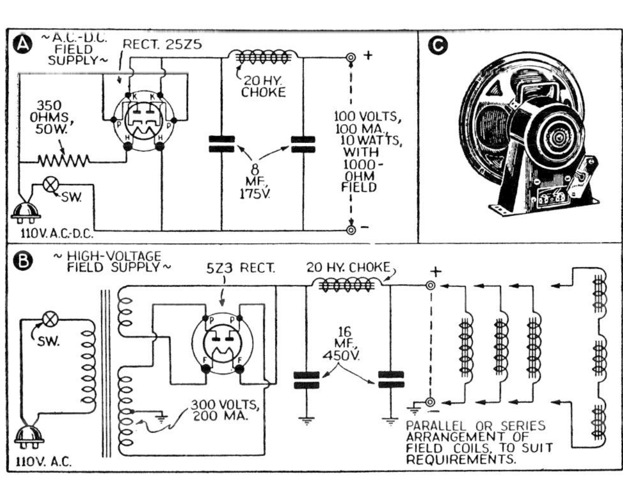

Fig. 19. Recommended speaker and optional power supplies.

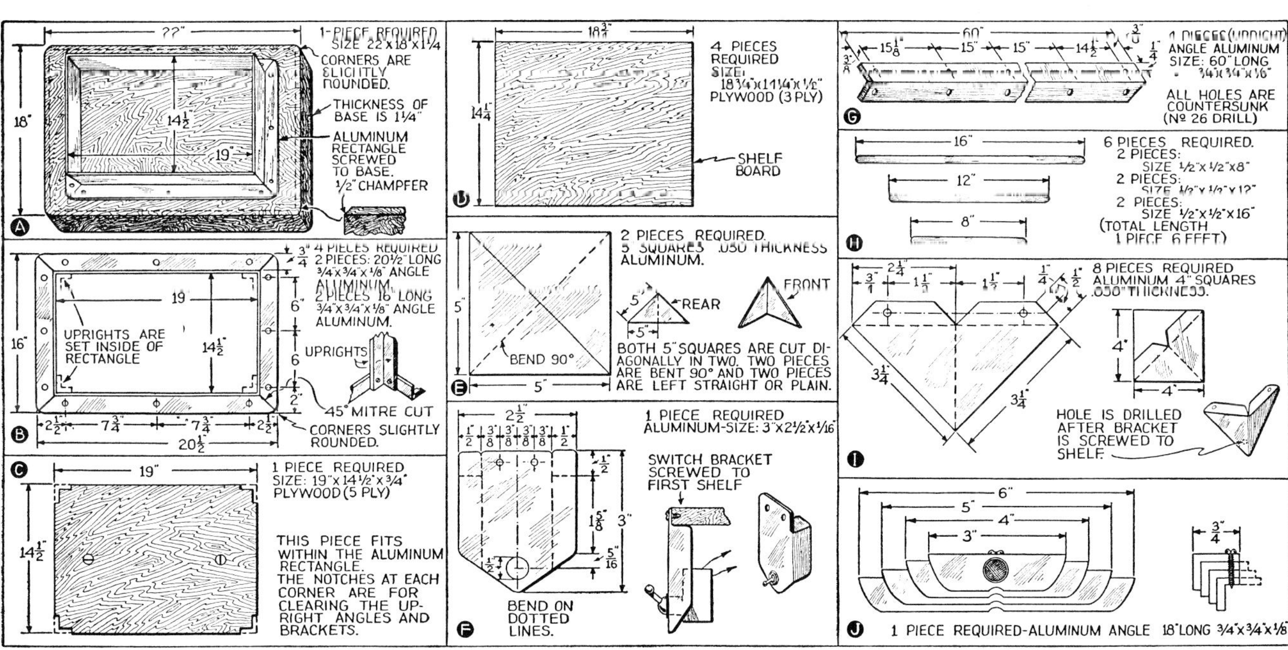

Fig. 18. Here are the steps necessary to construct the rack for the 4 chassis. These specifications must be followed exactly if the chassis are to fit.

Fig. A
XPD—"peanut" diode-triode for deaf aids. XVS—"peanut" variable-mu screen-grid tube.

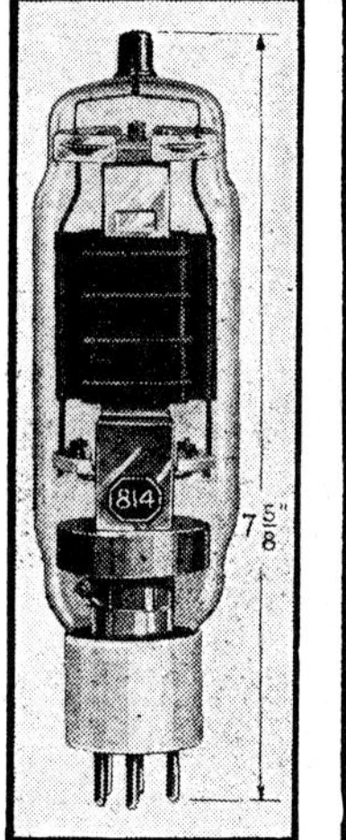

Fig B
814—transmitter R.F. and A.F. amplifier, frequency multiplier and oscillator (at 30 megacycles).

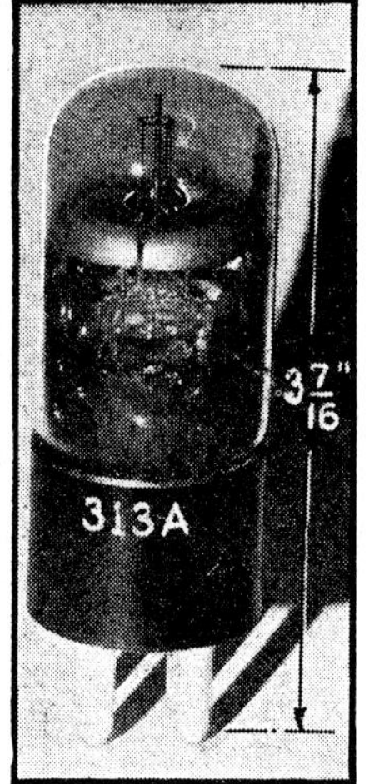

Fig. C
313A — cold-cathode gas-filled multi-purpose tube. It may be used as a relay, voltage regulator, or rectifier.

NEW TUBES

INTRODUCED TO RADIO MEN —IN 1937-'38

From infra-midget to ultra-giant, the 13 types presented this month cover a size and power gamut never before encompassed in one crop of new tubes. The intermediate types are of equal interest to many technicians.

R. D. WASHBURNE

IN THIS ARTICLE—

"Peanut" Diode-Triode
"Peanut" Variable-Mu Screen-Grid
Transmitter Beam Power Amplifier
Cold-Cathode Gas-Filled Relay, Voltage Regulator and Rectifier
250,000-Watt Transmitter Tube
Triode-Heptode Converter
High-Efficiency Rectifier
Gas-Triode Sweep-Circuit Oscillator and Grid-Control Rectifier
High-Efficiency Pentode Amplifier
"Carrier" Pentode Voltage Amplifier
"Carrier" Pentode Power Amplifier
Gas-type Voltage Regulator
Filament-type Current Rectifier

NEW demands upon the electronic field by various services have resulted in the development and production of several unusual tubes which we here describe; more orthodox but equally insistent requests from radio set makers for improved types also have been met by tube manufacturers whose latest designs are considered in this article.

XPD "Peanut" (2 V.) Diode-Triode. This tube—latest addition to the line of peanut-size types previously described in *Radio-Craft*—operates in conjunction with the type XVS tube; A.V.C. being fed back to the control-grid from the diode of the XPD (shown pictorially in Fig. A). The XVS being a variable-mu screen-grid tube it is thus possible to produce a deaf-aid having A.V.C. and capable of maintaining a constant-volume output level. This arrangement completely avoids blasting on loud transients. Curves are available from the English manufacturer. (*)

Characteristics data on this tube appear at the end of this article in Table I. Tube terminals are shown in Fig. 1. Note that either of 2 base types is available.

XVS "Peanut" (2 V.) Variable-Mu Screen-Grid. Although especially designed for use in deaf-aid apparatus this tube, also shown in Fig. A, due to its variable-mu characteristic may be recommended for use in small-space radio receivers in which it is desirable to have a variable-mu type tube in order to reduce cross-modulation. (*)

Characteristics data on this English tube appear in Table II. The pin connections are shown in Fig. 1.

814 Transmitter Beam-type R.F. Power Amplifier, Multiplier and Oscillator. This is a filament type of ceramic-base transmitter tube incorporating new design principles involving the use of directed electron beams. See Fig. B.

Features resulting from the use of these principles in the 814 are that the screen-grid absorbs little power and that efficient suppressor action is supplied by space-charge effects produced between the screen-grid and the plate; full power may be developed at 30 megacycles. The resultant high power-sensitivity makes this tube especially suited for use as an R.F. amplifier, frequency multiplier, oscillator, and plate-modulated amplifier.

Tentative characteristics and ratings, and maximum ratings and typical operating conditions as an R.F. power amplifier in class B telephony, are given in Table III. Further characteristics for class C telephony together with considerable operational data and curves are available from the manufacturer. Tube terminals are shown in Fig. 1.

(Data courtesy RCA Manufacturing Co., Inc.)

313A Cold-Cathode Gas-Filled (1) Relay, (2) Voltage Regulator, or (3) Rectifier Tube. The overall appearance of the 313A tube is shown in Fig. C. (In finished tubes the bulb is opaqued to render the glow invisible.) Due to lack of space *Radio-Craft* until now has been unable to present this most important contribution to the electronic field. However, and although not strictly a "new" type in comparison with the other tubes described in this article, it merits special attention in view of its application in the article, "Radio Waves Used to Completely Control Remote Receiver," which is appearing in April *Radio-Craft*.

As a circuit element, the 313A tube may be used to perform as a *relay*, as a *voltage regulator*, or as a *rectifier*. Within the tube there are two conduction paths, one between the 2 control electrodes, known as the *control gap;* and the other between one of the control electrodes and the anode, known as the *main gap*.

The 2 control electrodes are coated with an activated surface of barium and the envelope is filled to a pressure of several centimeters of mercury with a mixture of the rare gases, neon being the principal constituent. The physical properties of the barium surface together with the nature of the gas-filling combine to produce a discharge device that operates on an unusually low voltage.

Each gap is characterized by a breakdown voltage and a sustaining voltage. This difference in breakdown voltage of the 2 gaps enables the tube to act as a relay.

*Most Radio mail order houses can supply this tube if properly identified as to type, title of article, issue (month) of *Radio-Craft* and year.

Two features of the 313A in a suitable circuit arrangement are of particular importance. One is the flatness of the voltage curve for positive currents, which makes the tube suitable for voltage regulation, and the other is the lack of symmetry in voltage for positive and negative currents, which allows the tube to act as a rectifier.

Characteristics data for various operational circuits and conditions, and of performance curves, are available from the manufacturer; general characteristics data are given in Table IV. Tube terminal connections are shown in Fig. 1. This tube has been developed by Bell Telephone Laboratories.

Super-Power Transmitter Tube. Something quite unusual in transmitter tubes and one which is mentioned here to illustrate the tremendous advance in transmitter tubes is the 250-kilowatt giant-size tube illustrated in Fig. D. Heretofore high-power tubes have only been "man-size" but the new job now undergoing test at the Whippany, N. J., laboratory of Bell Telephone Co. is almost double the height of preceding types.

Inlet and outlet pipes feeding the outer jacket of this water cooled vacuum tube may be seen leading leftward. The upper or outlet portion of this cooling jacket is mounted in a suspension cradle that helps support the entire tube; the lower end of the tube is supported by a ring-mounting below which protrude the various connecting leads. Not shown in this view are the insulators which isolate the water-cooling system; and a foot-high corrugated insulator which both insulates and supports the weight of this tube. Vacuum tubes of this sort cost thousands of dollars.

6J8G Triode-Heptode Converter. This new converter tube (Fig. E) consists of a triode and a heptode unit having a common cathode and assembled in the same envelope. It combines a triode oscillator element with a heptode converter section, so designed as to avoid oscillator frequency drift with change in applied A.V.C. voltages.

Although the basing arrangement of this tube is such that in some applications it may be substituted for type 6A8G, with slight realigning, it is not primarily intended for use in this manner. As compared with 6A8G, the reduction of some of the interelectrode capacities is an important advantage in type 6J8G. High conversion gain is maintained at frequencies of 18 mc. and above, and the ratio of signal to noise is materially improved. The exceptionally high plate resistance of 4.5 megs. makes it possible to use a high-quality I.F. transformer with marked advantage in gain. The selectivity of a high-Q tuned circuit is not appreciably impaired by shunting the plate resistance of this tube across it. High input impedance is maintained under all normal operating conditions.

Characteristics data are given in Table V. Tube terminal connections are shown in Fig. 1.

(Data courtesy Hygrade Sylvania Corp.)

6ZY5-G High-Vacuum Full-Wave Rectifier. This tube (see Fig. E) is a full-wave rectifier of low heater-current rating, for use primarily in farm-radio receivers that obtain their power from a storage-battery vibrator supply. The tube is also well-suited for use in those farm receivers that are designed to operate from either a battery-vibrator supply or from an A.C. power line.

Tentative characteristics data for this tube are given in Table VI. Terminal connections are given in Fig. 1.

(Data courtesy RCA Manufacturing Co., Inc.)

884 Gas-Triode Sweep-Circuit Oscillator and Grid-Control Rectifier. Here is a tube which is destined to become exceedingly popular in cathode-ray equipment. Electrically the 884 is like the 885 except for heater rating. The 884 has a 6.3 V. heater which facilitates the use of this tube with other 6.3 V. tubes in the design of sweep-oscillator equipment.

Tentative characteristics data are given in Table VII. Terminal connections are shown in Fig. 1.

(Data courtesy RCA Manufacturing Co., Inc.)

6C6C Pentode High-Efficiency Power Amplifier. This tube (see Fig. E) is of the "low drain" heater series which requires 150 ma. at 6.3 V. It may be used

(Continued on page 621)

Fig. D. 1938's transmitter tube contribution to the electronic field, a 250,000-W. water-cooled giant.

Fig. E. Left to right: triode-heptode converter, gas-triode rectifier for C.-R. tube circuits, high-mu A.F. triode, carrier voltage amplifier, carrier power amplifier.

YOU MEN Who Buy RADIO!

By making your problems our problems, we know the service you require—and to render you such service the entire Allied organization is dedicated—to give you the service you have a right to expect. Tremendous stocks, remarkable values and a real desire to serve, all combine to make Allied your ideal source of supply.

Wholesale Prices

THIS big new catalog heralds a new era in radio. Never before have you been offered such variety—**such quality**—such sterling value in radio as you are now offered in this large new catalog. In it, **wholesale prices** are not a myth—but *actual* reality. You will marvel that such merchandise, comprising everything that is **new and worthwhile** in radio can be offered at such prices. Here you will find high quality that is not high priced. If you are interested in radio you cannot afford to be without the **Allied Catalog.**

SET BUILDERS!

Set Builders, Amateurs and so called "Hams" will delight in the *unusual variety*—and **remarkable values** that are offered in standard kits and parts. Tremendous stocks—**real organization**—prompt shipping service all combine to make Allied your ideal source of supply.

RADIO DEALERS!

The live radio dealer—the man who keeps pace with the rapid advance of radio will find much of **real interest** in the Allied Catalog. New A-C Sets, D-C Sets, Dynamic and Magnetic Speakers, television equipment, in fact **everything** that an impatient radio public is demanding.

What Allied offers you!

Allied Service will prove a revelation to you in what radio service can really be. Allied Executives backed by years of training in radio are practical men. They know radio. Their vast experience has built up around them an organization trained to serve. Months of effort have built up here a tremendous reserve of stock that makes for prompt shipments; and this stock is new stock comprising the seasons pick of such prominent manufacturers as Silver-Marshall, Tyrman, Aero, Hammerlund-Roberts, etc.

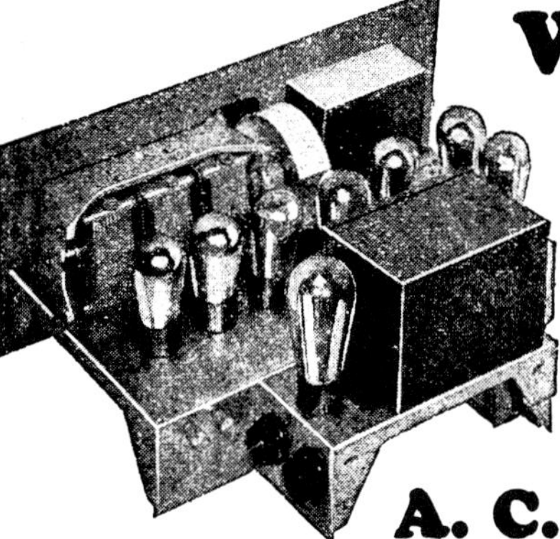

A. C. ELECTRIC SETS

Allied offers you a new—complete line of A-C Receivers, available in either chassis form or in a wide variety of beautiful console models. Prices range from $32.95 to $199.00. Dollar for dollar they stand out as one of the season's leading receivers. Engineered to unusual perfection they offer you features found only in the highest priced sets.

You Profit When You Buy Right

Buying right is half the battle. From the small set builder to the large dealer, your success depends upon gauging the public pulse of radio and in **buying right.** Everything that is new in radio—the items the radio public is now demanding are here, ready for your call. Write now—the catalog is **free** for the asking.

Write for Catalog Now

Allied Radio CORPORATION

711 W. LAKE ST. CHICAGO, ILL.

THIS ADVERTISEMENT APPEARED IN 1928

Times Change... Radio Sweeps On

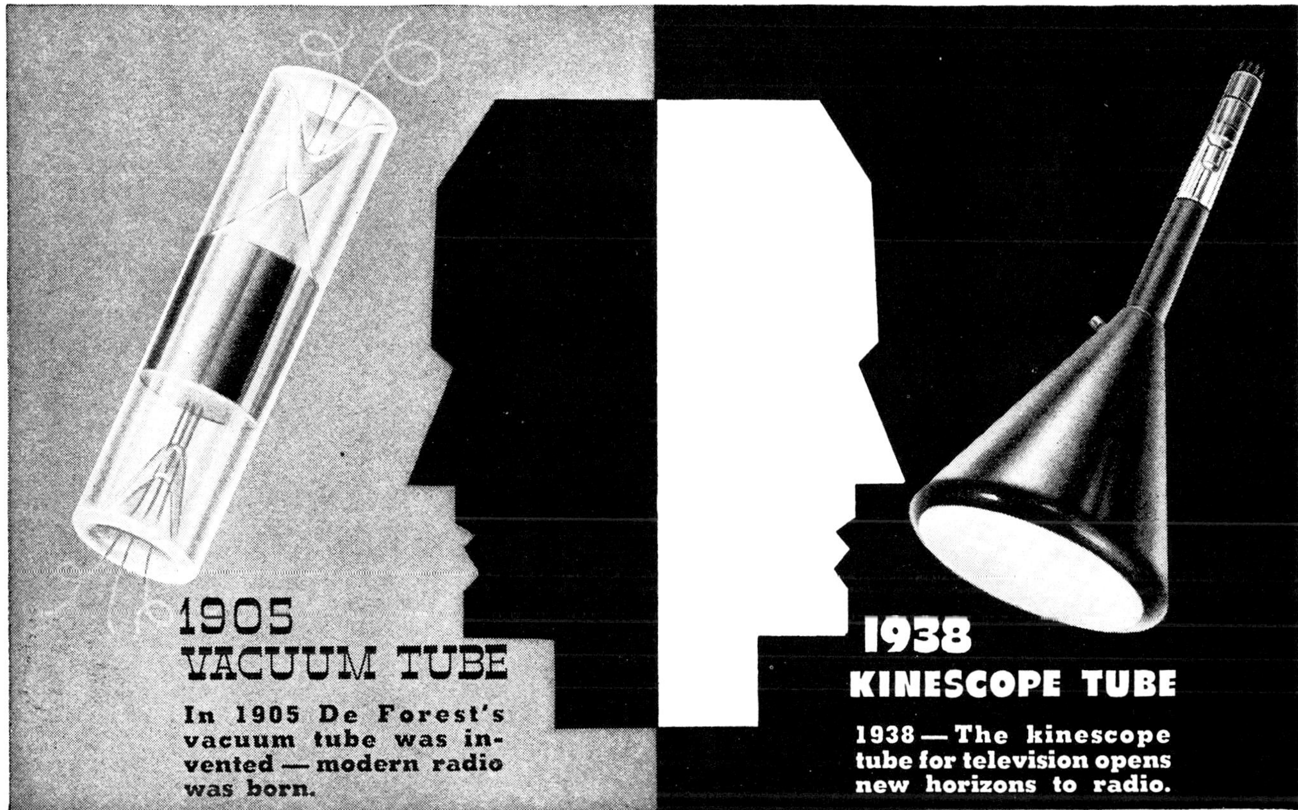

BUT Allied IS STILL THE LEADER!

To read the advertisement on the opposite page, which appeared in 1928, is to realize that ten years haven't changed ALLIED much — except to add to the thousands of radio men who find that they get what they want at ALLIED. Yes, today the merchandise is better. Our stocks are larger. We employ more people. And with our larger-volume purchases, we offer even greater values. But essentially, ALLIED is still the same — a house designed and built for service to the radio field.

We CAN'T change THAT — because that's what ALLIED IS, to thousands of service men, dealers, sound-specialists, radio-builders, and amateurs all over the world — the radio supply house that knows their needs and how to fill them, that gives them geared-up service and a "square deal" on every order, that serves as their headquarters for everything in radio, at lowest prices. That is ALLIED — 10 years ago — today — and 10 years hence.

If you haven't the latest ALLIED Catalog, write to ALLIED RADIO CORPORATION, Dept. 2-C8, 833 W. Jackson Blvd., Chicago, Ill., and we will send a free copy by return mail.

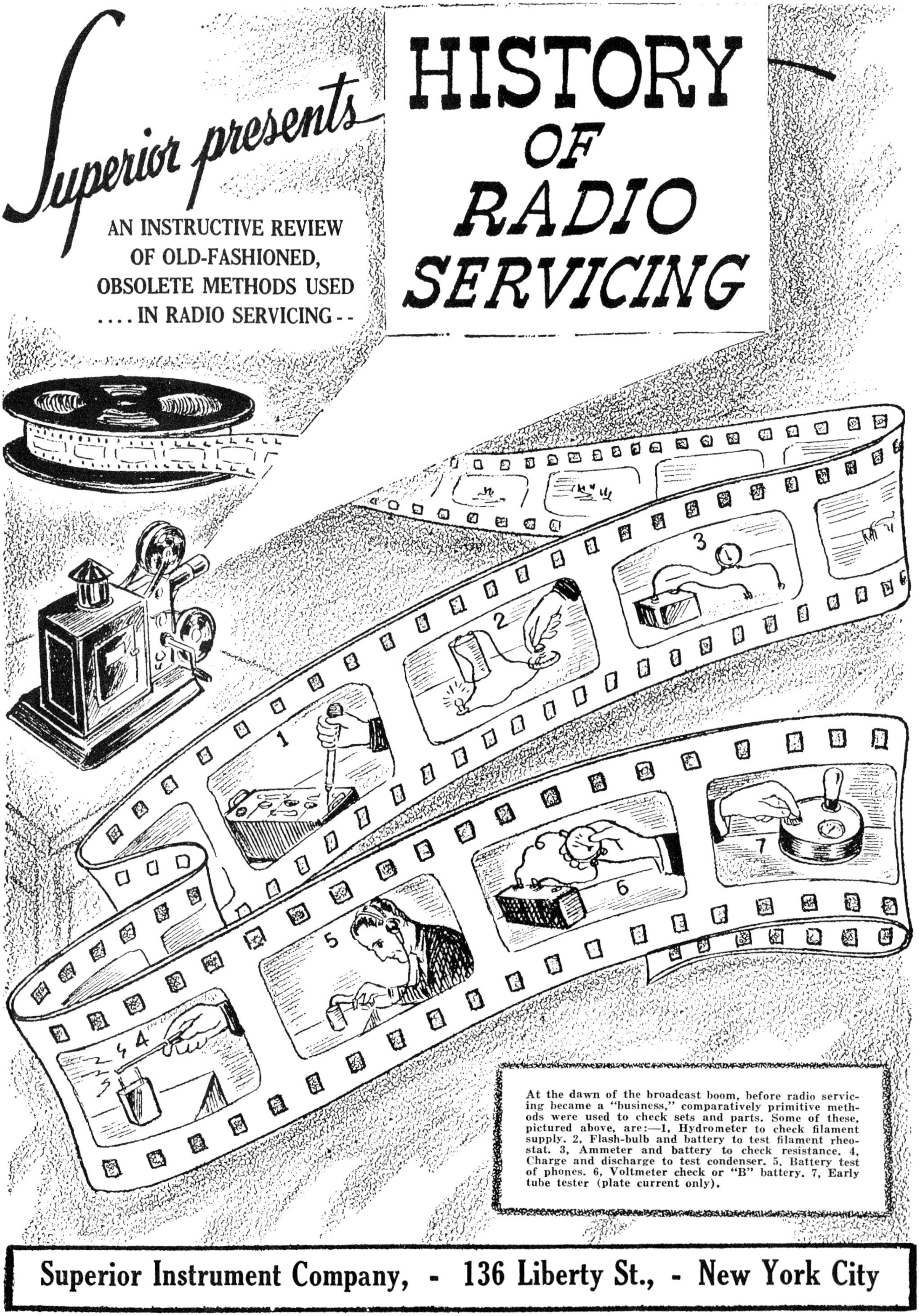

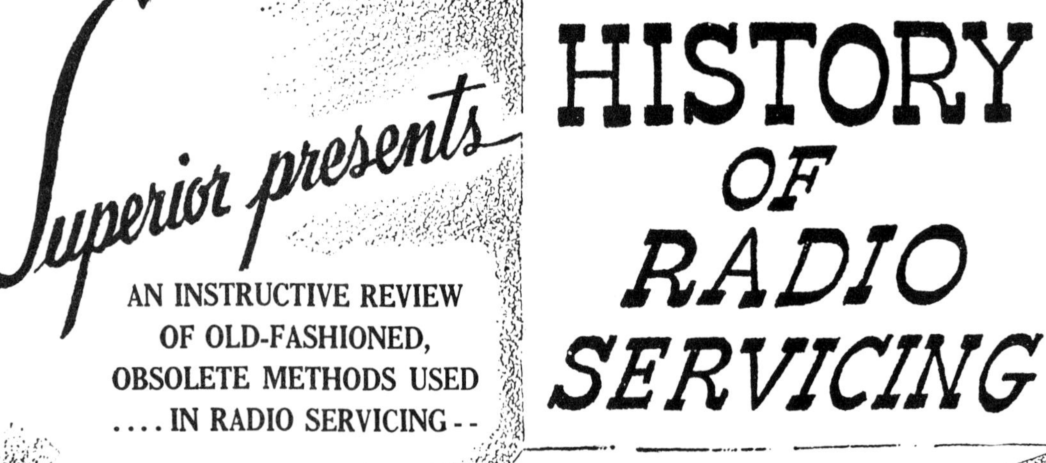

Superior presents
HISTORY OF RADIO SERVICING
AN INSTRUCTIVE REVIEW OF OLD-FASHIONED, OBSOLETE METHODS USED .... IN RADIO SERVICING--
1
2
3
4
5
6
7
At the dawn of the broadcast boom, before radio servicing became a "business," comparatively primitive methods were used to check sets and parts. Some of these, pictured above, are:—1, Hydrometer to check filament supply. 2, Flash-bulb and battery to test filament rheostat. 3, Ammeter and battery to check resistance. 4, Charge and discharge to test condenser. 5, Battery test of phones. 6, Voltmeter check or "B" battery. 7, Early tube tester (plate current only).
Superior Instrument Company, - 136 Liberty St., - New York City

An Oustanding FREE SOUVENIR IS OFFERED BY-SUPERIOR-TO READERS OF THE JUBILEE SOUVENIR NUMBER

A COMBINATION FOUNTAIN PEN AND AUTOMATIC PENCIL with YOUR NAME stamped in GOLD LETTERS, will be given FREE with ANY item purchased, as featured on this page.

Therefore, when ordering more than one instrument, we suggest that you send us additional names to be monogrammed. Please be sure to mention JUBILEE SOUVENIR NUMBER when ordering.

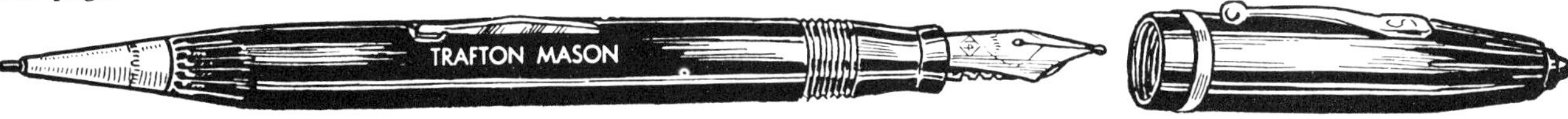

COMBINATION ALL-WAVE SIGNAL GENERATOR PLUS A COMPLETE TUBE TESTER

ONLY $19.90

• Truly the greatest value ever offered by any instrument manufacturer! There is no comparison to this COMBINATION INSTRUMENT, made by Superior, at the price of $19.90. Consists of two fully complete units housed in a single metal cabinet of attractive design: (1) an ALL-WAVE SIGNAL GENERATOR and (2) COMPLETE TUBE TESTER. • The Signal Generator has a range of from 110 kc. to 22 mc., all fundamental frequencies, in five bands—with front panel band-switching control. • All frequencies are directly calibrated on the dial, and a 4 to 1 planetary drive is incorporated to allow for slow precise tuning adjustment. • R.F. signal with or without audio modulation at the control of the operator. • Separate R.F. and A.F. outputs, with two A.F. amplitude levels. • Accuracy 1 per cent on I.F. bands, 2 per cent on short-waves. • Separate Neon leakage test circuit, for testing condensers, etc. • Employs two tubes; 2—6A7's and one special neon modulator tube. The TUBE TESTER tests all 4, 5, 6, 7 and octal base tubes, including diode tubes. • Incorporates separate neon short or leakage test, to determine whether tubes are bad, noisy or leaky. • English reading meter, with "BAD—?—GOOD" scale. • Uses minimum number of operating controls, allowing tube tests to be made quickly and accurately. • Sturdy, reliable and foolproof emission type circuit. • Operates on 110 volt A.C. current.

Shipping weight 19 lbs.

VOLT-OHM-MILLIAMMETER

The HANDI-METER, a new volt-ohm-milliammeter, enables direct reading of d.c. volts 0/5/50/500, also ohms 500/20,000, likewise direct-reading, and d.c. current determination to 20 milliamperes. Low ohms scale enables checking r.f. and i.f., as well as a.f. coils, for shorts and opens. Low-priced instruments never before included low-ohms range, reading to one ohm, nor did they have an adjuster for accuracy of ohms readings. Complete with test leads and self-contained batteries. Shipping weight 5 lbs. Net price **$4.10**

27 Instruments in 1

THE ALLMETER

1,000 OHMS PER VOLT

The ALLMETER permits both A.C. and D.C. measurements for volts and currents, also low resistances from below one ohm, also high resistance capacities, henries and decibels, **comprising 27 instruments in one.**

1—A.C. Currents, 15-150-750 ma. This extremely valuable service is practically never afforded by multitest instruments. Instruments, 3.
2—A.C. Volts, 15-150-750 volts. Affords also output meter service. Instruments, 3.
3—CAPACITY, .01-50 mfd., all condensers, including electrolytics. Instruments, 1.
4—5-1,000 Henries, coil loaded or not. Instruments, 1.
5—LOW ohms, .03-500 ohms. Extremely valuable. Instruments, 1.
6—HIGH ohms, to 0.5 meg., covering all practical values. Instruments, 1.
7—D.C. Volts, 15-150-750 volts. Simplified scale uses same calibration as for a.c. Instruments, 3.
8—D.C. Currents, 15-150-750 ma. The same scale simplification, also 0-1 milliamperes. Instruments, 4.
9—Decibels, –12 to +10. Instruments, 3 + 8 – + 30 (by interpolation) + 8 – + 50 by interpolation.
10—Vacuum-tube voltmeter, 0-15-150-750 volts A.C. D.C. Instruments, 6
11—Continuity tester. Instruments, 1. Shipping wt. 8 lbs. Net price **$10.40**

THE INFINOMETER

An Accurate V. T. Voltmeter

$14.40 *Complete*

• Measures Critical Voltages in Low-Current Circuits, Such as in Volume Expanders, A.V.C., A.F.C., Detectors, Oscillators, Noise-Suppression Stages, DC amplifiers, Photocell Circuits, Etc.
• Operates From 110 Volts DC or AC (any frequency), With Less Than 5% Error on High Voltage Ranges Due to Line-Voltage Fluctuations.
• Three DC Ranges; 0-6, 20 and 60 Volts: Four A.C. Ranges; 0-5, 15-50 and 150 Volts. Sensitivity approximately 700,000 Ohms per Volt on DC Ranges; Approximately 300,000 Ohms per Volt on AC ranges.

Shipping weight 8 lbs.

THE RC BOX

SOLVES PUZZLING PROBLEMS IN CAPACITY AND RESISTANCE REPLACEMENTS; ALSO USEFUL AS STANDARDS OF CAPACITY AND RESISTANCE.

SPECIFICATIONS

Direct-reading, 0-40 ohms, at 10 watts maximum allowable power dissipation • Direct-reading, 0-20,000 ohms, at 3 watts maximum allowable power dissipation • Direct-reading, 0-1,000,000 ohms (1 meg.), at 1 watt maximum allowable power dissipation • Direct-reading, 10-100 mmfd. (.00001 to .0001 mfd.), at 200 peak volts allowable maximum • Direct-reading, 100-450 mmfd. (.0001-.00045 mfd.), at 200 peak volts allowable maximum • Rheostats of specially selected material for retention of calibrated values, including wire-winding of the 40 and 20,000-ohm units • Tuning condensers of exceptional quality (excellent power factor) • Two separate binding posts for each of the six different services, all clearly designated. (Rotor of two-gang condenser is common to two posts, upper left and upper right) • Accuracy is 2 per cent for all capacity settings, also for all settings of the 40 and 20,000-ohm resistors; 5 per cent for the 1,000,000-ohm unit • Instrument is totally shielded, so that spurious effects due to stray pickups, affecting capacity measurements particularly, are avoided. One of the common rotor posts may be used as ground when the other condenser or any of the resistors is being used. Overall size of the instrument is 10x6½x4 inches • Black wrinkle finish cabinet has carrying handle. Net price **$6.10**

Shipping weight 6 lbs.

GENEMETER

WITH VARIABLE AUDIO FREQUENCIES

$14.40

Complete with Four Tubes

• Combination R.F. and Audio Signal Generator, R.F. 100 KC. to 105 M.C. A.F. 25-10,000 cycles, all direct reading, all by front panel switching.
• R.F. and A.F. outputs independently obtainable alone, or with A.F. (any frequency) modulating R.F.
• Output meter for connection across primary of receivers' output transformer for peaking with modulation "on."
• R.F. is subject to attenuation, and oscillation leakage is minimum.
• Condenser and other leakages tested to 100 megohms.
• Main dial protracted on 7½" diameter, and 4-to-1 vernier planetary drive.
• Accuracy 1% on I.F. and broadcast, 2% on higher frequencies.
• All services on 90-130 volts a.c. or d.c.

Shipping weight 10 lbs.

MODEL V-45 TUBE TESTER

Emission type perfected tube tester for any and all type receiver tubes, including metal and metal-glass. Tests shorts and leakage. Accurate, simple, sturdy, direct — English — reading (Bad-Good). **$10.40**

Operates on 110 volts, 60 cycle, A.C.

Shipping weight 9 lbs.

SUPERIOR INSTRUMENTS COMPANY

136 LIBERTY STREET DEPT. RJ NEW YORK, N. Y.

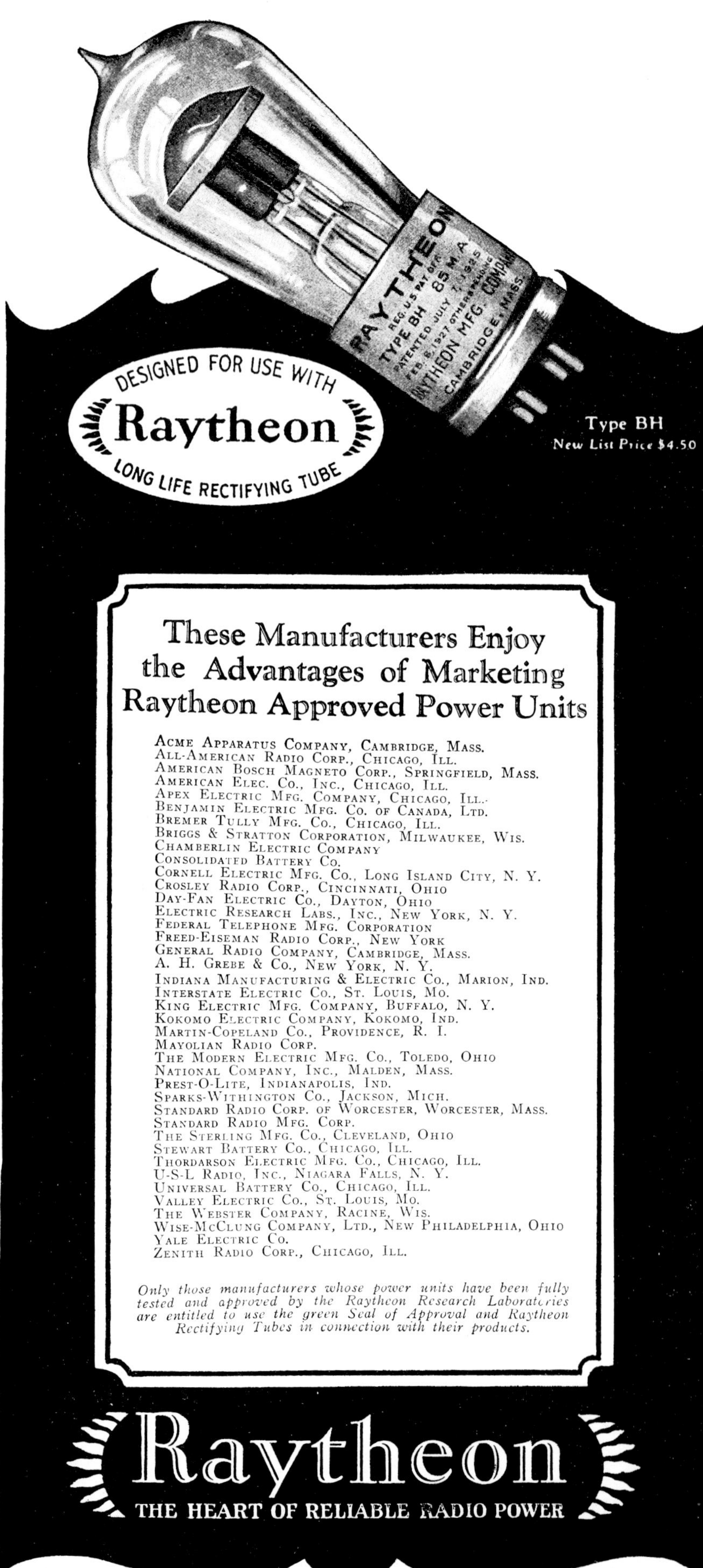

THIS ADVERTISEMENT APPEARED IN 1927

THE GROWTH OF BROADCASTING

(Continued from page 545)

scarcely going below 200 or over 2,500 cycles. This resulted in loss of "highs" and "lows" in the transmission of music and lack of natural quality in speech; the only reason it was acceptable is that early receivers were incapable of reproducing much more. As receiver design improved, the telephone companies kept step, offering radio lines of ever-increasing fidelity. Today almost every line for program transmission covers at least 100 to 5,000 cycles.

CAME SUPER-POWER—AND HIGH FIDELITY

Of approximately equal importance was the development of super-power for radio broadcasting. Early stations had radiated low power, in most instances 100 to 500 watts. Considering the larger figure, one is immediately struck by the fact that this power, being dissipated in all directions, is slightly less than that consumed in the average 1-slice electric toaster.

With so little energy being radiated, the ratio of static to signal was extremely high; therefore static disturbances drowned out received signals very often during reception. But station engineers battled against interference, both natural and man-made, the fight culminating on May 2, 1934, when Powel Crosley, Jr. (who had begun broadcasting in 1921) increased WLW's power to half a million watts (in the antenna system), the first station of that power in the United States (or, for that matter, in the world).

High fidelity, about the same time, was receiving attention and it was found that carrier waves could be modulated to carry audio frequencies from 30 to 14,000 cycles, representing the upper and lower limits of average hearing. It was found also that stations could transmit such programs without incommensurate expenditures, and experimental high-fidelity transmission was begun. Only a few of the better receivers were able to take advantage of this improvement, however, until the manufacturing branch of the industry took cognizance of the better transmissions and began producing receivers capable of reproducing the extreme highs and lows. WOR claims to be the first regular commercial station to have adopted high-fidelity transmission as standard practice.

IMPORTANCE OF SHORT WAVES IN BROADCASTING

Many consider the short-wave link as being next in importance. It has made possible not only the "stunt" broadcasts from airplanes, submarines, golf links, and similar points impossible to "hook-up" otherwise, but also the more important foreign programs. It antedates the previously mentioned steps by more than a decade, N.B.C. having used it in the mid-'twenties for the presentation of European programs in America, at not infrequent intervals.

Other advances are, for the most part, somewhat less revolutionary. For example, WENR-WMAQ claims a "first" on the broadcast of football and baseball games, and on the presentation of school and college classes by radio.

Other stations have originated other programs of similarly great importance, but immeasurably as such attention to programming has benefited radio, there is insufficient space to pay to each advance the tribute it merits.

MILE POSTS IN TELEVISION

(Continued from page 579)

The images, reproduced by cathode-ray tubes are clear, with good detail and adequate brilliance. Size is still unsatisfactory for general use, although the newer tubes, which produce 8 x 10 inch images, are more nearly adequate, though there is some loss of definition, which is more apparent than real.

It must be stressed that the comparison of number of lines per television image with number of dots per inch in a printer's halftone cut is unfair to television. While the imperfections of the human eye are such that it corrects for a certain amount of lack of detail in either still or moving images, it does so in a much greater degree when image motion is present.

Intermediate-film and projection-type television have been aided by new high-intensity illuminants; and, as recently pointed out in *Radio-Craft*, high-definition video transmissions may soon utilize up to 1,000 lines/frame.

Already in acceptable form for public presentation, the release of television may now be expected almost from month to month. There are a number of stations throughout the United States broadcasting television programs on an experimental basis; television is well established in England (where it is subsidized by the government) and upon the European continent. The major networks are ready for television they have studios and mobile (truck) pickups constructed—and are in truth straining at the leash.

With sufficient business upturn, to make mass purchasers of television receivers in the $200 class probable, broadcasting will commence. As to who will bear the cost of programs—sponsors will, despite the comparatively few sets which will comprise the early audience, for there will be publicity and prestige for the first advertisers in the new art.

The contention that the Federal Communications Commission does not permit televised advertising will no longer exist, for it is safe to say that this important government bureau will do nothing to hamper a budding industry, but will cooperate and try to aid it in every way.

Thus, with television, an old catchword will gain new meaning. It is:— "I'll be seeing you!"

DEVELOPMENT OF SOUND EQUIPMENT

(Continued from page 571)

baffle design, such as the Stromberg-Carlson "labyrinth", the RCA "acoustic chamber", the Philco "inclined sounding board" and the Emerson mantle set "sonic chamber".

As there is an increasing tendency to combine radio receivers and phonographs in a single cabinet, there must be special mention of radio's debt to the phonograph, and vice versa. At first the debt was all in the phonograph's favor. The needle of an Edison ("hill-and-dale")-type soundbox, resting on the diaphragm of a headphone, "amplified" the sound by means of the phonograph horn. Later, as has been described, an adapter was made to couple the phone directly to the phonograph tone-arm, and still later special speaker units were made, to replace the soundbox. Rhamstine, years ago, brought out a diaphragmless unit with a vibrating armature; the needle in the soundbox of a phonograph of the Victor ("wavy groove") type was then rested in a groove in the armature. (Incidentally, other vibrating-reed reproducers, including Clyde Fitch's "Ensco" unit, were made to utilize all sorts of things—from a piano sounding board to a door-panel—as their diaphragms.)

The phonograph had undergone a far longer period of development than the loudspeaker, hence the trend was to make use of the acoustic advantages of its reproducer system.

But radio developed apace. Soon the amplifiers and loudspeakers equalled the tone of the phonograph—then surpassed it. Magnetic pickups, essentially reed-type speaker units with a needle holder on the reed, came into use. An early pickup put out by Pacent had no arm of its own, but clamped to the tone-arm of the phonograph which nearly every family then possessed.

Pickup design has had many variations, but has never abandoned the early basis, though a collateral branch has come into being. There have been, and still are, magnetic pickups with the armatures damped by oil or rubber, but they still adhere to the same principle that an armature moving in a magnetic field generates an alternating current of frequencies corresponding to its vibrations.

The other type in general use is the crystal pickup, utilizing the current-generating properties of the piezo-electric crystal when subjected to mechanical stress.

Any of the pickups, when connected to the detector or first-audio stage of a receiver, either directly or through an impedance-matching transformer, permit the reproduction of disc-recorded music through the loudspeaker.

Many are sold in conjunction with electric phonograph motors and turntables, but these are a separate field and save for mention that the combination of radio and phonograph was probably responsible for the replacement of the spring motor by the electric motor, they need not be considered here.

Sound-on-film is also a distinctly related industry, not included although directly attributable to radio which developed the necessary amplifiers and speakers. Most interesting is a reference in *Modern Electrics*, dated May, 1912, for in that issue there is described the first sound-on-film talking motion picture apparatus. Though the light-beam method had not then been devised, the rudiments of the idea were there. But in this case it was proposed to have a "hill-and-dale" sound track on the edge of the film.

Hence, we can see that the present highly-developed art of sound equipment and reproduction owes much to radio. Its fostering of an appreciation of music and good reproduction, then development of apparatus to satisfy the demand for more and better programs, has done as much for the public as has the public done for radio. By no means are the sound engineers of today content to rest on present laurels,—as is indicated by the almost daily release of new developments from the sound laboratories. Even 3rd-dimensional sound has been developed and is now being perfected in the laboratories,—to produce realism along with high-fidelity reproduction. Some day in the near future we shall be "seeing" as well as listening, and the sounds will be related to the motion that takes place so that in every detail we will get an exact impression of being "Johnny on the spot" instead of alongside a radio receiver.

NEW CIRCUITS IN MODERN RADIO RECEIVERS

(Continued from page 595)

to the 2nd-detector has no gain and by disconnecting the diode there is no distortion of the signal from the pick-up due to rectification of it by the diode.

(5) SPLIT-LOAD PHASE INVERTER

Fada Model 316. A very practical application of the split-load type of phase inverter feeding 4 tubes in push-pull-parallel is shown in Fig. 1E. The plate load of the 6C5 A.F. amplifier is divided into two equal sections, one being placed in the cathode side of the plate circuit and the other in the plate side in the conventional manner. In this way, instead of all of the signal *voltage variation* appearing at the plate, half of it appears at the plate and half at the cathode. The signal voltage-drops *in the two loads* are of opposite phase and of equal value so that the following grids may be appropriately coupled to these points. Resistor (6) is the usual grid bias resistor and the total of this with resistor (20) makes up the cathode load resistance. One pair of power amplifier grids is fed from the plate of this phase inverter and the other from the cathode.

Radio-Craft considers this department to be an especially useful addition to practical radio literature. Comments from *Radio-Craft* readers will be appreciated.—*Editor*

THIS ADVERTISEMENT APPEARED IN 1912

PREDICTIONS AND FULFILLMENTS

(Continued from page 594)

like. There will be an entirely new industry built right around the radiotrola in 1922 . . . . The radiophone will come into more general use on the automobile, the motor-boat, the yacht and the airplane."

Every one of these predictions was borne out by subsequent events. Unfortunately, the next prediction, that of March, 1922, was likewise true, as the industry learned to its sorrow, for Gernsback wrote, "Present conditions will last for at least another year and the radio business will be a very healthy one for several years to come . . . . We will then witness . . . . price cutting, when real competition will begin in earnest. This condition, however, is at least six months distant . . . ."

But the art as well as the industry attracted the attention of this radio clairvoyant, and in the April-May, 1922, issue, Mr. Gernsback foresaw that, "We would not be surprised soon to have special copyright notices printed on all new musical selections somewhat as follows: 'This selection must not be used over the radiophone without special permission of the owners.' We can see where the composers will derive the same swollen incomes from the radiophone as they derive from the phonograph records." Ask ASCAP about that one!

Horn attachments were used with our radio sets in 1922, and Mr. Gernsback in the October issue, suggested using larger diaphragms than the 2-inch ones then prevalent, stressing the need for greater area in the reproduction of music. He also deplored the use of metallic horns. The paper-cone speakers which followed a few years later were the answer.

In November, 1922, he raised the question of adequate instruction sheets to be supplied with all receivers as, at that time, most of them were sold without such printed information. Today every set carries data covering its installation and operation. That is an important prediction, for the wider use of instruction sheets has greatly increased public satisfaction with radio by virtue of improved performance when correctly operated.

In February of 1923, Mr. Gernsback wrote, "We predict . . . . that radio frequency amplification will be used much more generally . . . . In 1923 we shall, no doubt, see the single-control outfit . . . . The next great improvement needed . . . . is the outfit . . . . that will do away with interference . . . . Only an expert can tune out the unwanted station . . . . The next improvement . . . . is the loud talker . . . . We predict that it will not be of the horn type at all . . . ." These predictions were followed by the neutrodyne which afforded more selectivity and soon was followed by ganged condensers. The cone speaker came somewhat later.

Wider use of educational programs was predicted in the April edition, and today such programs constitute an essential part of many daily broadcasts.

The year's final issue carried a Gernsback editorial which again mentioned the need for simplified tuning, and went into considerable detail about what should be done. The point about the undesirability of the horn was made again, but this time Mr. Gernsback added a radically new idea, "That loud speaker, by the way, will be built into the future radio cabinet, and why shouldn't it be?" It has been, as well you know.

It was in March of the following year, 1924, that he first predicted automatic tuning. It was similar to present push-button sets, except that Mr. Gernsback thought of jacks rather than buttons, and did not contemplate the addition of manual tuning in the same model.

June, 1924, saw a Gernsback prediction of radio installations in road houses and country inns. He asked, editorially, "If city restaurants install radio loudspeakers, is there any reason why those

in the country cannot do likewise? . . . . A loud speaker in a store or inn will surely attract people by the hundreds." Now we know who originated the idea which has made soothing silence a lost boon in America!

"Television," wrote Mr. Gernsback in January, 1925, "will come to the front during the next three years." And this statement was promptly followed by the booms of Jenkins, Baird and others. And a little over a year later, in February, 1926, he predicted a television and servicing instrument, the cathode-ray tube (used in an oscilloscope), saying, "Some day some one is going to invent a new sort of lens by which it will be possible to look into a radio set and actually see the radio frequency currents . . . ."

Another tube, the screen-grid, then made in Europe though not commercially available in the United States, was predicted in August, 1926. Through their use, he wrote, "it should be possible to produce amplifiers of power far excelling anything we have hitherto seen." Today's R.F. and I.F. amplifiers are proof that again he was right.

Sharper tuning was called-for in a March, 1927, editorial, and soon was made available by the manufacturers, who apparently watch Mr. Gernsback's every utterance with an eagle eye.

June saw another Gernsback editorial on television. In this he told of the need for motors which would stay in synchronism. A year or so later, synchronous motors appeared on the market, not only for television, but in midget form for clocks as well.

He continued his discussion of television needs in the December, 1927, edition, saying, "I personally foresee the final television receiver, which will have incorporated into it some sort of vacuum tube, such as the Braun. The Braun tube, instead of using discs, or wheels, employs a cathode discharge. This latter, having no inertia, throws upon the screen a moving beam of light which responds to magnetic impulses. There should not be any difficulty in getting this beam to work at any required speed." In view of present trends in television, all that one may comment is:—!!!!!

In May of the following year, he pursued the subject farther, prognosticating that early television sets would replace not only the loudspeaker, but would use the rest of the equipment, and that subsequently they would be built into the sound radio console, operating from house current. This, too, has happened.

The same editorial predicted the smaller console and the midget table model as a distinct trend. This, similarly, came about.

September, 1928, brought forth another bull's-eye prediction when Mr. Gernsback wrote, "It would not surprise me at all if, during the next five years, the broadcasting of both sight and sound will be done completely on short waves . . . ." This was borne out by fact. The Columbia Broadcasting System accomplished it first, early in 1932.

Turning to some other of the writings of Hugo Gernsback, we find an editorial in the October-November, 1931, issue of *Short Wave Craft*. In this, Mr. Gernsback discusses the use of police radio and the possibility that criminals might listen-in to the alarms dealing with them. To obviate this possibility, he suggests that code words be used so that even if unauthorized persons do listen, they can learn nothing. Since that was written such code signals have won very nearly universal adoption throughout the United States.

In the March, 1933, issue of the same publication, Gernsback stresses the need for better band-spread in short-wave tuning, and for "noise-free" antennas. It was only a little later that the wave of enthusiasm for "split-second" tuning and for doublet antennas with twisted-pair lead-ins was manifested.

The need for work on the waves lying below 10 meters was considered in an editorial in the same publication for December, 1933, and the comparative freedom from static on such waves was pointed out.—As you know, television in the United States has turned entirely to these ultra-highfrequencies.

Turning to *Science & Invention*, another Gernsback publication, we find more about television, at an even earlier date, for in the August, 1921, edition, Mr. Gernsback editorialized: "Within a comparatively short time . . . . we will be able to witness the future Dempsey-

(Continued on following page)

Again by Sprague

Since the first electric car in 1888—a Sprague invention—many electrical engineering advancements have been given industry—by Sprague.

No wonder that the latest—the SPRAGUE MIDGET fixed condenser—is attracting such widespread set-builder interest.

For the MIDGETS are factory tested at 1500 volts—have higher uniform breakdowns—constant capacities—insulated casings—quarter the weight—half the size; in short, they are midgets with electrical hearts of giants. Their new flexible lugs—presoldered—short-cut assembling expenses; saving screws, nuts, bolts, drillings and wire, too.

Just your address will bring you samples and all the reasons why Sprague Midgets will improve and economize in *your* set construction.

SPRAGUE SPECIALTIES COMPANY
Dept R.
QUINCY, MASS.

THIS ADVERTISEMENT APPEARED IN 1927

PREDICTIONS AND FULFILLMENTS

(Continued from preceding page)

Carpentier fight a thousand miles away with our own eyes, by radio, while out at sea. In other words, television by radio will be an accomplished fact." The truth of that statement is apparent, in the light of progress which has already been made.

The same publication, in March, 1924, ran another typical Gernsback prediction. He wrote, "At the present time there is a slow swing from the telephone head set to the loud speaker. Local dealers will tell you that the proportion of headphones sold to loud speakers is 5 to 2. But in another year or two, the proportion of the loud speakers will probably be larger than that of the telephone receivers."

An odd branch of radio, still little known to the layman though familiar to physicians, was predicted in the October, 1924, issue, when Mr. Gernsback editorially discussed possible effects of radio waves on the human body, expressing the belief that they would be beneficial. It was not until nearly a decade later that short-wave therapy came into use, however.

In September, 1927, Mr. Gernsback made a double prediction. He wrote of the coming of television, stating that within 20 years it would have won general adoption. This prediction is cited, although the period has not elapsed, to show that Mr. Gernsback was not one of the crowd who, each year, confidently predicted television "next season". His other prediction forecast air conditioning—an industry of no little inerest to many radio manufacturers, distributors and Service Men.

He turned his attention to wired pictures and the trans-Atlantic telephone in the November, 1927, issue, stating that " . . . . when the rates are reduced sufficiently to make it attractive, people will use the service. And, incidentally, the inventions will go begging no longer, but will coin handsome dividends." The telephone rate from New York to London was $25 a minute in those days; it is now $21 for 3 minutes during business hours and $15 on Sundays. Traffic has increased tremendously. Similarly, the wire-photo service is in daily use by newspaper syndicates.

Lack of space prevents more complete discussion of the forecasts of Hugo Gernsback, but the collator wishes to add that there is hardly an issue of any technical or popular scientific publication which Mr. Gernsback has owned or edited that does not contain one or more predictions. A large percentage of such predictions has come true. The fact that not all of the Gernsback statements have as yet been proven in no way casts doubt upon their accuracy. The wireless transmission of power, interplanetary communication, and telemechanics are well within the bounds of possibility. Work has already been commenced in these fields.

Today's dream becomes tomorrow's fact.

BOOK REVIEW

THE RADIO AMATEUR'S HANDBOOK (1938 Edition), by the Headquarters Staff of the American Radio Relay League. Published by the American Radio Relay League. Size, 6½ x 9½ ins., paper covers, 564 pages. Price, $1.00.

This book offers even more comprehensive treatment of the amateur short-wave field than the 14 previous editions. It contains about 600 illustrations, 290,000 words, 73 charts and tables, and 111 practical equations and formulas. It is, perhaps, the largest volume of technical information to be sold at so low a cost, and represents 4 months' work by 12 specialists in amateur radio.

Among the new chapters are one on workshop practice, another on portable and emergency equipment, and another on fundamental principles. The latter is written for the complete novice.

All other chapters have been completely rewritten and modernized. Every radio man should have this reference work available.

RADIO IN THE FUTURE

(*Continued from page* 591)

school children would find this a much more pleasant method of doing much of their homework.

The physical size of radio sets, particularly when it comes to portables, is undergoing radical change even now. Already it is no trick for the short-wave enthusiast to build a radio set that will bring programs from other hemispheres, which set can conveniently be placed in one's coat pocket. Future radio is going much further than this.

Radio tubes are continually being improved. They are also getting smaller and smaller. We already have radio tubes with a diameter no larger than the little finger. The day is not far distant when an efficient radio tube will have shrunk to the size of a rubber eraser at the end of a lead pencil. The future tube will probably be "cold"; that is, instead of having a glowing filament, electrons will be supplied by some chemical or other means. That means, of course, that the power which we require of a radio set will be infinitesimal. I foresee the time when the pocket set will not require any batteries at all; whatever electricity is required in order to operate the set will be supplied by a tiny electric generator housed in a case no larger than your watch. Indeed, the generator will be driven by means of a large watchspring, and will run for a considerable period. At present there is already manufactured a small flashlight, hand-operated by means of a tiny dynamo, but the bulb consumes a comparatively large amount of power.

So the watchspring-driven electric generator becomes a certainty in the not too distant future. In fact, the entire radio set will be small enough so that the watchspring generator, molecular speaker and whatever tubes are necessary will go easily into the vest pocket. Indeed, the set will be small enough so that it can be concealed in the hand. You will say, as a toy this might be interesting, but what practical use is there for sets of such type? The answer to this is that as civilization progresses, instant communication is of paramount importance. Twenty years ago you probably would have laughed at the idea of car-radio, but in the future no one will wish to be without his vest-pocket set because after all, we are not always within reach of the more efficient, large radio sets. We may be in the street, we may be on a train, we may be in a theatre, in a restaurant or any of hundreds of other places, and may wish to be in touch with the outside world.

By that time, of course, the radio set will not only be equipped for broadcasting but for personal messages as well. Physicians, contractors, business men, and others find it necessary to be in touch with their offices many hours of the day. Special broadcast stations will be erected whereby special messages to certain individuals will be broadcast. This is particularly important for physicians. Such broadcast stations probably will operate at special frequencies. In other words, this will be a glorified paging system, but not limited to the confines of a hotel; it will be possible to page individuals within a radius of 100 miles.

By this time every radio set, whether pocket or otherwise, will be television equipped, but this is no longer news. You may ask, how will I be able to see a television program by means of a vest-pocket set? The answer to this is found in another prediction, not as yet realized, which I made in the October, 1936, issue of *Short Wave Craft*. This has to do with an instrument which I term television "spectacles." In other words, you wear a pair of "spectacles" which have a twin television receiver built right on the eye-glass frame. This means, of course, that the television images received are small, but they need not be large, because they are right up close to your eye; indeed they will be sharper in this manner. Therefore, all you require is to put on "television eyeglasses" and you will see and hear at the same time, because the frame of the television set on the bony part of your nose will also, by molecular vibration, transmit the sound to your ear-drums in a most realistic fashion.

And let no one think that this exhausts the list of future radio developments. As I have said many times before, *the really big radio inventions are still in the future.*

THIS ADVERTISEMENT APPEARED IN 1926

SUPER-REGENERATION—IN 1922

HERE is an illustration of Major Edwin H. Armstrong's super-regenerative loop receiver—about 1922. The original caption speaks for itself: "An invention that amplifies an ordinary radio receiving set 100,000 times and eliminates the use of outside aerials is the latest work of Major Edwin H. Armstrong, formerly in charge of Uncle Sam's radio service in France. This super-regenerative circuit with 3 vacuum tubes gives the same result as an ordinary receiving set would give with 9 vacuum tubes. Another feature of Major Armstrong's invention is its simplicity—it having but 3 adjustments." The essential super-audible or quenching frequency was obtained from an oscillatory circuit that included the 1,250-turn honeycomb coil which appears so prominently (alongside the box under Andrew Halbran's left hand) in the photo. The photo-shy major couldn't be induced to pose with his achievements of "making 3 tubes do the work of 9," but he did snap this picture.

PROGRESS OF RADIO RECEIVING EQUIPMENT

(Continued from page 550)

usually employed Edison or nickel cells, made in test-tubes and housed in an open rack which was provided with a switch to connect blocks of them in parallel for charging. Filling 60 or more of these cells with distilled water at regular intervals was no small bother, but the cost, though initially high, was lower in the long run than for drycell "B" batteries.

Eventually the eliminators appeared. These were at first used only on the audio stages, the detector requiring smoother current than was possible with rectification and insufficiently effective filters.

Such eliminators were almost a necessity when one owned a "big" 5-tube set, which fairly ate up batteries—until the 201A type tube was put on the market. As the 201's used 1 ampere each, the ¼-ampere 201A's afforded great economy, as far as the "A" drain was concerned, and were bought up faster than they could be produced. Many of the more conservative listeners, however, held out to the last, buying 201's instead, considering them more sensitive, "better tone." The same thing happened years later, when the 171 was replaced by the 171A, which many thought to have a lower output.

The 5-tube sets were, at first, simple *tuned-radio-frequency*, or "T.R.F.", jobs. T.R.F. was a Navy patent, and it was released about the mid-twenties, when almost all the manufacturers rushed to use it. Torn between the desire to put out an effective set and the wish to produce one economical of operation, many of the engineers bent their energies to the production of *reflex* sets, in which the currents were fed through the tubes at radio frequencies, detected, and then put through the same tubes again at audio frequencies, each tube doing the work of two. The Acme, Sleeper and Radio Guild sets were splendid examples of this type.

Reflex sets used untuned-R.F., for the most part, simply tuning a loop antenna with a condenser across the grid circuit of the 1st R.F. stage, interstage coupling being effected through iron-core R.F. transformers. Having but one tuned circuit, they were not especially selective. The ads said, "The . . . . set gets all the stations" and the users added, "Yeah, but at the same time!"

Nor were the simple, straight T.R.F. sets an unalloyed joy. If well made, they squealed almost as badly as the bloopers; if inefficient, they would not get as much DX as the users wanted—and the lust for DX was well-nigh uncontrollable. To rectify this, manufacturers began introducing "lossers" of various types. One of the earliest and best was a potentiometer in the R.F. grid-returns, to vary the bias on these stages, in order that the tubes might be operated just below the point of oscillation. Other manufacturers put fixed or variable resistors in the R.F. grid leads, while still others advertised that the trick was done with interstage shielding—and put the shields in the fields of the R.F. coils to introduce loss in that way.

Then came the *neutrodynes!* They were the sensation of radio. If you didn't have a neutrodyne, you were such a social outcast that your very dog hid under the bed at your approach and looked at you reprovingly, because the dogs of neutrodyne owners wouldn't associate with him.

The neutrodyne used a low-capacity trimmer condenser of unique design between the grid and plate of each tube in the R.F. to neutralize the internal capacity between these elements. The Hazeltine corporation, which owned the patents, formed its licensees into an organization that advertised as a group and protected its patents mutually.

Besides its high efficiency and freedom from oscillation, the neutrodyne had a novel and valuable feature—you could actually "log" it! That is, a station often would come in at the same place on the dials (there were 3 of them) twice in succession. "Daddy," said the little boy in the advertisement, "get Los Angeles." "Just turn the dials to 32 and there it is," the father replied — and there it was indeed, especially if the set was in Hollywood.

The *3-circuit regenerative* sets were of about the same period. These were the old "bloopers" with an untuned primary antenna winding loosely coupled to the grid coil. Theyblooped, but the oscillations did not radiate, and a degree of peace came to the kilocycles. These sets, too, could be logged after a fashion.

Then came the "supers"—the *superheterodynes.* There had been a few super-regeneratives, but they were mostly home-made; the superhets. were factory produced. RCA put out some really good ones, with the I.F. stages sealed in a mysterious can (known as the *catacomb*) to prevent tampering.

One of their best sets at that time used a new and wonderfully economical tube—the UV199—which required only six one-hundredths of an ampere filament current, and had a remarkably low plate current as well. It could be operated from small batteries, which could be housed in the cabinet, and neatness of installation took a large stride forward.

One of these sets, put out in the middle '20's, had a sloping panel, much like many present sets, and operated on a loop antenna which projected from its lid. The cabinet could be had provided with legs—and console models could then be seen dimly in the offing. There was also a portable super., which had its loop concealed in the cover, much like several of the portable T.R.F. and reflex sets. This one, the model 26, was a really first-rate job.

It was about then that "A," "B" and "C" eliminators came into use, and electric sets were foreshadowed. Indeed, some sets which used special tubes that would operate with A.C. on their filaments made their appearance, but they met with considerable skepticism. Fans,

(*Continued on following page*)

When the Neutrodyne made its Bow!

Professor Hazeltine (about 1923) demonstrating his famous Neutrodyne. Said the press, "One of the most startling developments that may revolutionize radio is this . . . 5-tube Neutrodyne receiver which is incapable of oscillating, is non-radiating and utilizes 2 stages of tuned radio frequency, detector, and 2 stages of audio frequency . . . can be operated entirely by dry batteries." The coils' "sacred angle" of 54.7° seems also to be holding the interest of J.D.R. Freed (pointing), who built the set.

THIS RADIO CORP. OF AMERICA ADVERTISEMENT WAS PUBLISHED IN 1921

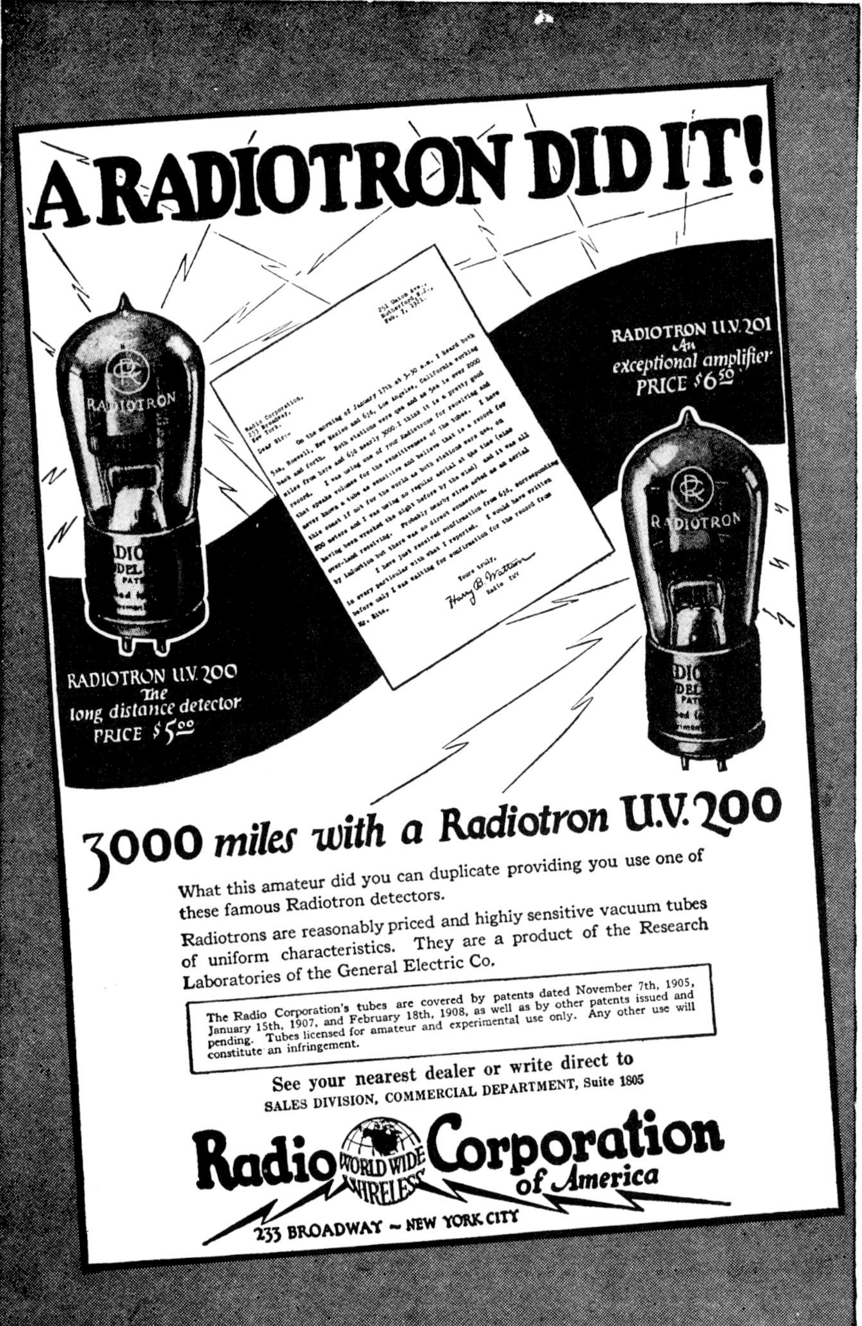

THIS ADVERTISEMENT APPEARED IN 1921

Early Tube Experiments

As soon as the de Forest Audion made its appearance inventors throughout the world immediately started to manufacture tubes of their own, often with the idea in mind to circumvent the patents.

Other combinations were tried in order to improve if possible the original Audion.

One such attempt of an early vacuum tube that would not only not infringe on the patents but which would also have claims and points of its own was the Gernsback audion-crystal detector combination (shown at right). This odd tube was a 3-element tube similar to the de Forest Audion. It had the usual filament and the usual grid but instead of the plate it had a crystal. In this particular tube there was a piece of silicon forming the plate of the third element of the tube. Mr. Gernsback thus combined the crystal detector and vacuum tube. Of course the device worked, but perhaps not better than the original Audion. Only a few of the tubes were manufactured for research purposes, the tube itself was never marketed.

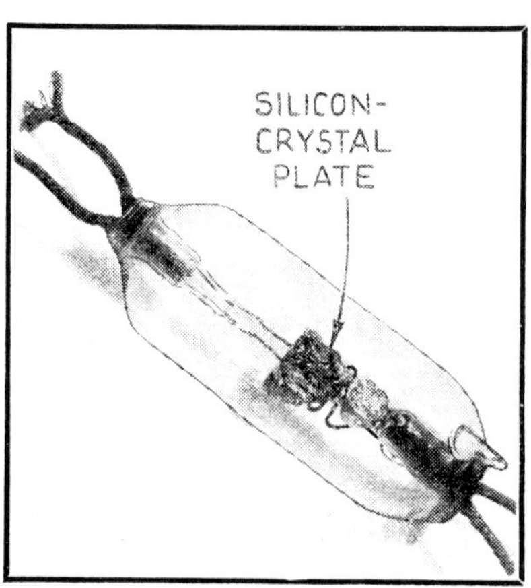

(Continued from preceding page)

who had been given a separate rheostat for each tube, felt that there simply would be enough hum to interfere with DX reception, and the sets did not meet with the popularity they merited.

Late in the '20's another new tube arrived on the radio scene—the screen-grid tube — an even greater advance than the power output tubes of some years before. The latter had made it possible to increase the bass response of the set; the former simplified R.F. amplification, for they were efficient, yet could not oscillate. With their advent, the neutrodyne circuit began to wane in popularity, as sets using '22's became the fashion.

A short time afterward, the '27's again revolutionized radio. These were the first tubes that the public could be convinced would work well on A.C.—with no "A" eliminator needed. They heralded the birth of the "all-electric" set—radio as we know it today—for while the 210's and other A.C. power tubes had been used, people had still been rather dubious as to the wisdom of trying A.C. in the earlier stages. Soon thereafter the '24's—A.C. screen-grid tubes—were out, and radio went to town. The fashion was A.C. T.R.F.

Superhets. had not enjoyed the fullest possible popularity because, in most of them, not only did the dials fail to "track," or read the same for a given station, but in many instances the same station could be tuned-in at two points on the dial—and listeners had learned to demand "accurate logging."

It was not until the early 1930's that single-control "one-spot" supers. (A.C. operated, of course) began winning the lead which they have never since lost.

Tone control was the next real achievement, and while it has since won universal acceptance, it met with a mixed welcome at first, as some set users held that it merely added another knob to turn. These good people, who now thought 3 controls too many, were probably the same people who had happily twirled the 17 knobs and dials which tuned the 3-tube sets of a decade before.

Tuning meters, so that the stations could be brought in "on the nose" were another useful adjunct of the same date. These were, at first, simply milliammeters in the screen-grid circuits of the I.F. tubes, when automatic volume control was used. This A.V.C. was, by the way, a distinct advance, for it eliminated the blasting which had formerly shattered the eardrums when a powerful station was tuned-in after a weaker one. The tuning meters were, with one or two notable exceptions, merely meters, the popular "shadow" type simply substituting a vane in the path of a light beam for the familiar needle. The cathode-ray type, popularly called by the copyrighted (RCA) name "Magic Eye," did not come into general use until the middle '30's.

Quiet automatic volume control, "Q.A.V.C.", which smothered inter-station background noises by cutting sensitivity when no carrier was being received, began attracting notice about 1934 and has since won general acceptance, and the following year, all-wave tuning became something that every set had to have, save in the lowest-priced models. The folks who previously had the honeycomb-coil sets sat around with looks of fond recollection on their faces. But the new sets were far simpler to operate; band switches (remember the "Navy" sets?) made shifting from one range to another merely a matter of turning a switch, to cut one group of coils out of the circuit and another group in.

Then came high fidelity. Early sets had reproduced audio frequencies from about 200 to 3,000 cycles, then had improved to take from 100 to 3,500 cycles. For a while it was believed that 100 was about the lower limit, and that 60-cycle A.C. hum would come through if the sets went much lower. But as filtering was improved, some de luxe models of the early '30's handled 30 to 4,500 cycles with no trouble. By '35, high-fidelity sets reproduced from 10 to, in some instances, 14,000 cycles, resulting in far more natural speech and music.

Automatic frequency control, which moved mysteriously to make the set tune perfectly when tuned approximately by hand, came next. This resulted in the re-birth of automatic tuning, which had been tried and virtually abandoned some 10 years earlier.

Old-time sets had used both mechanical and motor-driven automatic tuners, but the difficulty of compensating for minor tuning variations had rendered such systems somewhat less than 100% satisfactory. Now A.F.C., as well as automatic temperature compensation (obtained through the use of thermostatic trimmers), have made automatic tuning practical.

Radio has become respectable. Sets are complete units, with self-contained speakers, as Hugo Gernsback prophesied when such perfection was undreamed. But is the radio listener of today as happy as the fan of yesteryear? I doubt it, for though programs have improved along with equipment, the old thrill is largely gone. No more does one wonder, every time the set is switched on, whether it will actually work! Perfection has replaced chance in radio reception.

BOOK REVIEWS

OLD WIRES AND NEW WAVES, by Alvin F. Harlow. Published by D. Appleton-Century Co. Size, 6 x 9 ins., cloth covers, 548 pages, over 70 illustrations. Price, $5.00.

Here is a zest for chronology of communications dating from the days spoken of in Homer's *Iliad* down the years to the present day. Throughout the book is the note of optimism that spurred on the indefatigable workers of yesteryear.

Every effort has been made to maintain authenticity; and where the author found historical statements confusing, he presented, to the best of his ability, both sides of the story and allowed the reader to judge.

The average radio man may feel that this book is incomplete in not treating extensively with the subjects of broadcasting and television, and other phases of radio, however, it must be remembered that the book is intended only to present the historical background of *communications*. In this connection the book is quite complete in discussing point-to-point telegraphy and telephony by wire and radio. The undeniably important part that amateur radio played in communications development, however, is not mentioned; and, too-great emphasis is placed upon the telephone.

Taken all-in-all, though, the volume makes interesting reading; and is an excellent reference to historical dates in communication. It will be a valuable addition to any library.

MICROPHONE MEMOIRS, by Credo Fitch Harris. Published by The Bobbs-Merrill Company. Size, 5½ x 8¼ ins., cloth covers, 281 pages. Price, $1.75.

This book is light reading of the most entertaining sort. It is dated for the period subsequent to 1922; its reading tempo is timed to suit the dilettante who wants to really enjoy an extensive reminiscent period. Although the locale revolves around WHAS, Louisville, Ky., the retrospection is quite representative of the more glamorous days of radio's infanthood—or to quote the author, "the Horse and Buggy Days of Radio."

The charming informalities of Radio's early period are gone forever, and the odd and amusing customs of the past can be found only in such works as this.

THIS ADVERTISEMENT APPEARED IN 1925

IN APRIL RADIO-CRAFT (ANNUAL ELECTRONICS NUMBER)

BENJAMIN F. MIESSNER—Greatest living authority on electronic musical instruments, tells *Radio-Craft* readers about "Latest Developments in Electronic Music."

GLENN H. BROWNING—Co-originator of the famous Browning-Drake receiver, starts a series of articles on an ultra-modern, easily-built all-wave kit set.

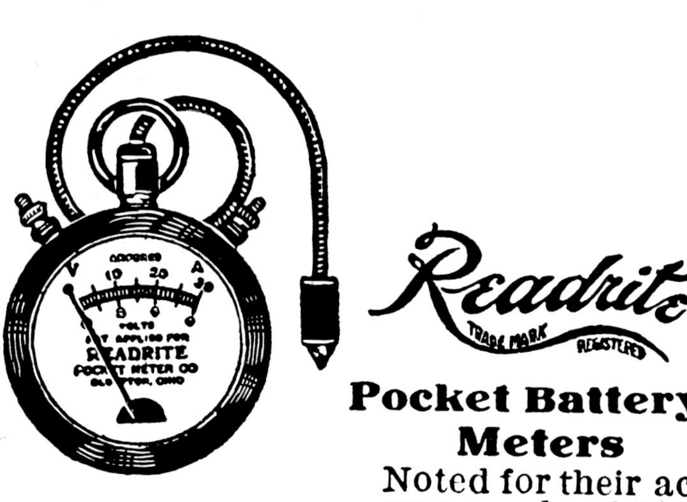

Pocket Battery Meters

Noted for their accuracy, durability and permanency. A written guarantee for ONE YEAR supplied with each meter.

Ammeters, $2.50 Voltmeters, $3.00
Volt-Ammeters, $3.50 and $4.00

Write for circular and discount.

Read-Rite Meter Works, Mfrs.
14 Main St., Bluffton, Ohio

THIS ADVERTISEMENT APPEARED IN 1908

THE STORY OF AMATEUR RADIO

(*Continued from page* 563)

United States was trying to maintain its neutrality. Government officials had reason to suspect the Telefunken station, WSL, at Sayville, L. I., and asked amateur C. E. Apgar, 2MN, whose work was known to them, to help check. He recorded the German station's transmissions and turned them over to the Secret Service. The transmissions proved to be reports on allied shipping, sent to German submarines, and WSL was closed and confiscated.

The United States and Germany were drifting closer to war, and in the autumn of 1915, the A.R.R.L. told the government its members would cooperate if the occasion arose, and was cordially thanked. Then in April, 1917, all amateur stations were ordered closed by the United States government.

Shortly after our entry into the war, the Navy called upon the A.R.R.L. for 500 capable operators and any good radio equipment that might be available. The request was complied with inside of 10 days, and by the end of the war, more than 3,500 A.R.R.L. members were in the service as operators.

The amateurs were allowed to recommence operation of their receivers on April 12, 1919. On September 26 of the same year all bans were called off, transmission was permitted again, and before the end of the year, amateur radio was re-established.

With experience gained in the war-torn years, the amateurs began replacing their spark transmitters with tube oscillators (C.W.), and a few of them then started to send out voice and music in addition to code. One such amateur was 8XK; you can read more about him in the article "The Growth of Broadcasting" (in this issue), for he was Frank Conrad, whose pioneering in radio transmission of music gave birth to the present billion-dollar radio entertainment industry.*

In the latter part of 1921, Paul F. Godley was sent to Scotland to erect an amateur listening post, so that trans-Atlantic tests might be conducted. So many were the American amateur stations that he heard, and preponderantly C.-W. stations, too, that the prestige of the amateur rose immeasurably as a result of that test. The fact that the amateur could do so much with so little power and on wavelengths that were previously held in contempt was a revelation to both the public and the giants of the industry.

The amateurs—all 60,000 of them now operating in America today—are more than a mere part of the radio picture. They are a part of life in the United States. An important part, too, for they continue, year after year, to serve the nation faithfully, whatever be its needs.

*Dr. Lee de Forest, Frank King, and George Eltz, as well as the *Detroit News* had staged earlier musical broadcasts, without important result.

SOME OLD-TIME RADIO STATIONS

(*Continued from page 589*)

spectrum. No small wonder that these stations had to use tremendous power and extremely large and lengthy antenna systems. The tuning coils on the receivers, for receiving these wavelengths, were also very large—sometimes resembling huge Tesla induction coils. In general, though, commercial wireless seemed to favor the 600- to 1,000-meter range.

The first attempt at legislation of commercial wireless was made at the International Radiotelegraphic Convention, held at Berlin in 1906, which provided that the wavelengths of 300 and 600 meters be assigned for commercial wireless work. The wavelength of 1,600 meters (or band from 600 to 1,600 meters) was reserved for governmental coast stations. However, these provisions did not at that time apply to wireless in the United States, since this country sent no representative to the Convention. The U. S. had no real laws governing wireless until The Wireless Law of 1910 was signed by President Taft, and which only became effective on July 1st, 1911. The tenets of the International Convention were not ratified by the Senate until January 22nd, 1913.

From this period on, and until the United States entered the World War, commercial wireless grew by leaps and bounds, and when finally this country did enter the War, the shore stations that were established by this time proved to be an invaluable asset. When the War started there were 35 Navy coastal stations (privately-owned commercial stations were much more numerous), and 45 Marconi stations for ship-to-shore service. These the U. S. Navy took over, and in 1918 purchased them outright. Their calls were heard by innumerable wireless operators of those days, and amateurs as well, since listening-in to commercial news, press, etc.—especially, the NAA press reports after the 9.55-10.00 P.M. time signals—was a favorite pastime that provided good practice to attain "copying speed."

Today, the number of shore stations in the United States is difficult to estimate, since wireless (radio) is no longer confined to ship service. Aviation, Police, Facsimile, Press, Communication (radiograms), Wireless Telephone, Broadcast, Military (Army and Navy), Experimental (television, etc.), all tend to make the number run into many thousands.

Meanwhile the ghosts of old stations still linger in the background. Arlington still sends out Time Signals; Nauen, Germany, still transmits on the upper wavelengths where now, if one were to listen-in, an almost graveyard silence prevails in contrast to the old-time busy chatter. The Eiffel Tower is somewhat more modern, having included facilities for radiophone and, we understand, television transmission, but the call letters are the same . . . probably to serve as a reminder of its past glories.

To Readers of RADIO-CRAFT

FREE POSTCARDS MAY BE HAD UPON WRITING TO THE PUBLISHERS

These post cards make it easy to answer advertisements which appear in RADIO-CRAFT, without cutting valuable data which you may wish to save.

Many times manufacturers request you to "clip the coupon" when answering their ads. Often this means destroying part of an article on the reverse page you may need later for reference. Save your RADIO-CRAFT issues complete. If you should ever want to sell bound volumes, or certain copies of RADIO-CRAFT, the resale value of **uncut** issues is very much higher than that of mutilated ones. **So send for a supply of these free post cards and use them in answering all RADIO-CRAFT advertisers.**

REMINISCENCES OF OLD-TIMERS

(Continued from page 561)

of the Institute of Radio Engineers on his typewriter. The Institute of Radio Engineers was young and broke, so Doc carried on. Later Vice-President in charge of Engineering for RCA. An inspiring force—with a mind loaded with engineering data seasoned with a priceless sense of humor—to the younger men.

Curtis, back from erecting the wireless station at Manaōs, the modern city now buried in the jungle. What a yarn he spun.

Armstrong, new on the horizon with his mystery black box, filled the phones with signals. Armstrong later, over a flying field in France, changing a ballast lamp on a wind-driven generator far out on the wing.

Fessenden, the burly giant, whose genius produced steel antenna towers, the heterodyne, the radio-frequency alternator, and the Arlington Station. Fessenden was an addict to the smoking of his own special brand of poisonous, black, skinny, long and particularly vile stogies. No survivor ever thereafter accepted another offering of them. Fessenden delegated Bill Beakes to press the key, for the first time, at the opening of Arlington. Bill tied 2 broomsticks together—for distance seemed desirable—and gave it a tap. A deafening crash. Bill dropped the stick and made a World's record 50-meter sprint. Fessenden has passed on, and today Beakes heads-up the United Fruit Co. radio system.

Hazeltine, starting in radio by giving our crowd in Navy Radio a much needed lesson in tube mathematics, and winding up as one of the great names in the field.

Coffin, ladling out stiff doses of Bjkernies in German as part of his job, and creating for his own amusement, his vector analysis, and the formulas we all use for calculating inductance.

Clark, over at the Navy, sticking pins into turns and measuring inductance for his coil-curve library. An athiest who refused to believe the Bureau of Standards' "Inductance Bible".

Wiegant and Ferrand, battling at the Institute over the static eliminator.

Simon, sparkling with enthusiasm, surrounded by assorted genius, producing exotic masterpieces of design, and pioneering in aircraft radio, and radio intercity communication networks.

Eaton, keying high powers, and bringing sanity by his modesty and clear thinking to the Washington Navy Yard madhouse.

Loftin, running around in France with 8 Eiffel towers under one arm and a 1,000-kilowatt arc job under the other, later to erect Lafayette, for many years the most powerful station in the world. Afterwards he was to tackle the job of studying and evaluating every patent in the radio art for Uncle Sam.

Alexanderson, of the agile sliderule and contempt for mere size. He gave us Gargantuan antenna systems, and transmitters. From paper-thickness steel he evolved ponderous machinery that safely whirls at the edge of destruction, spaced by microscopic clearances maintained by one of his brain children.

Fuller and Beal, provided the arcs that proved that Navy operators can drink the food of the generators of their design, namely, alcohol mixed with benzine; and live, and come back for more.

Jack Binns, and he makes me think of the lonely memorial in Battery Park, inscribed with the names of the heroes of wireless, true men who maintained the traditions of the sea.

There are many more. The pathfinder team of **Stone Stone** and **Roos.** The famous **Cutting** and **Washington** team. **Craft** and his Army mule test. **Logwood** the inspired. The logical **Ballantine. Squires** with his waves on wires. **White,** his tubes and his common sense. **Godley** and **Maxim,** egging-on the "hams" with their distance work. **Cooper-Hewett** in the Madison Square Tower. **Davis** the empire builder. **Lowenstein** the perfect engineer, and later on, the KDKA group.

De Forest, the master magician, I have saved for the pièce de résistance. The creator of 5 billions of dollars of American industries. His inventions are the essential starting point of much of the present-day discoveries. Thus far we can credit him with the invention of the vacuum tube, broadcasting, the long-distance telephone, the commercial radio telephone, motion pictures with sound-on-film, the radio knife, high-frequency therapeutic fever equipment, and a host of elements of these. Television would be impossible, the photoelectric cell would be still a laboratory curiosity, the concentric cable undreamed-of without his inventions. You who have prospered from his inventions, or those whose lives have been saved by his radio knife, or who have had death-dealing germs removed from their bodies by his fever equipment, as well as the uncounted millions whose spirits have been lifted and strengthened by the influence of his radio broadcasting and talking motion pictures owe this man an incalculable debt. Fortunately this brave soul is with us today, a tireless worker and unsoured by the fact that his magnificent contributions have not lined his pocketbook. (America has produced only one other giant of his measure, the beloved **Edison.**) Gentleman, our hats off to the greatest member of our craft!

WILLIAM DUBILIER—

Vice-Pres., Cornell-Dubilier Electric Corp.

PIONEER in radio and specialist in condenser design and manufacture. His activities in this field are responsible for the many improvements in condensers since the era of the Leyden jar. Holds innumerable radio patents pertaining to telegraphy, telephony, high-frequency apparatus, and a system for detecting submarines which he devised during the World War.

I FIRST became interested in radio when I read in the local papers that Marconi was coming to this country to lecture on his wireless telegraphy apparatus in 1903.

In 1904 and 1905 I assisted in giving lectures for some of Marconi's associates who remained here and then began to build some of the early coherers and open spark coil transmitters, using medical induction coils.

In 1906 I was associated with the Continental Wireless Telegraph and Telephone Company.

It is not generally known that I operated the equivalent of a broadcasting station as early as 1909. The novelty of receiving music through the air appealed to the owner of an amusement park in Seattle. He fitted up a crude receiving set and erected a sign "Listen to the Wireless for 10c." Strangely enough, every time that I visited the amusement park and attempted to listen-in, I was told that the apparatus was out of order. The receiver was working, but there was no music to pick up—for the simple reason that I was not at the transmitter to operate it!

In 1908 I formed my own company—the Commercial Wireless Telegraph & Telephone Co. I remember buying equipment in 1908 or '09 from the Electro Importing Co., run by a young man named Hugo Gernsback who had a room upstairs on West Broadway and who was selling wireless apparatus, Branley coherers, spark coils and other equipment.

In 1907 and 1908 I exhibited radio, telephone and telegraph apparatus at the electrical show at Madison Square Garden, then at Madison Avenue and 23rd Street, for the Collins Wireless & Telephone Company. About the same time, I also organized one of the first amateur societies with a membership approximating 350. The first meeting was held in Montclair, New Jersey. During those early days of radio, it is strange to recall how the industry started with patent litigation and seems to have continued so up to the present time. I definitely recall the constant patent fights between the Marconi Company and the United Wireless Telegraph Company. Mr. Samuel Darby, father of the

(Continued on page 622)

THESE ADVERTISEMENTS APPEARED 1925-6-7

NEW TUBES INTRODUCED TO RADIO MEN —IN 1937-38

(Continued from page 601)

to advantage in home-type battery sets where low heater consumption is desired; with the same plate voltage it has about 4 times the power sensitivity of the 38, which requires twice the heater current.

The 6G6G is a pentode-type power amplifier tube designed for service in the output stage of receivers where maximum over-all efficiency is required.

Characteristics data are given in Table VIII. Terminal connections are shown in Fig. 1.

(Data courtesy Raytheon Production Corp.)

310A Pentode Voltage Amplifier for Carrier Systems. Increasing use of *wired radio* makes it necessary that the progressive technician keep himself posted on new developments in this field. It is of interest therefore to note the features of several new tubes especially designed for use in such carrier-current service. The 310A indirect-heater tube for instance, shown in Fig. E, has the following operating conditions.

The control-grid normally operates at 3 V. negative with respect to the cathode. The screen-grid of the tube provides electrostatic shielding between the control-grid and the plate. This is accomplished by *externally* tying the screen-grid to the cathode through a condenser, thereby effectively grounding this grid at voice and carrier frequencies. The 310 has about 3 times the voltage gain of the type 102F tube previously used in this carrier service.

Partial characteristics data are given in Table IX; terminal connections are shown in Fig. 1. Bell Telephone Laboratories developed this tube.

328A. Except for its filament characteristics this tube is a counterpart of the type 310A described above. Filament rating: 0.425-A. at 7.5 V.

311A Pentode Power Amplifier for Carrier Systems. The power output of the 311A (Fig. E) is about 9 times greater than that of the older, type 104D tube which is the power tube used in existing carrier systems.

The 3rd-harmonic is about 35 db. greater in the new tubes but this is permissible since the new carrier systems use feedback and the 3rd-harmonic is thereby reduced to an acceptable value.

Unlike tubes of this general type commonly available for radio receiver use these new tubes are designed to meet the much more severe demands of carrier "telephone" service. For instance, the insulators are supplied with a ceramic material to further improve the insulation, and the small metal plates used for electrostatic shielding are formed to act as baffles to prevent metal, sprayed from the surface of the cathode, from being deposited on the insulators and causing leaks.

(Continued on page 623)

REMINISCENCES OF OLD-TIMERS

(*Continued from page* 620)

present Mr. Darby, was the then famous patent attorney. (No wonder the present Samuel Darby knows his onions.) The de Forest Company sued the Collins Wireless Company and Fessenden sued them all. I recall being present at the various cases when textbooks and ordinary fiction were read into the testimony, one side trying to air out the other by making the suit expensive. It is strange to relate that some of the litigation relating to patents of that date were only recently settled by the Court of Claims—the Marconi patents claim against the Government.

During those old days, it was a constant sales struggle among the various companies to see who could sell more stock. Marconi Company started with a small office of 2 rooms in the Lords Court Building on William Street. The original stock of this company skyrocketed from $5 to $125 before they had a station going. The various companies started by de Forest sold many millions of stock to the public, all the companies subsequently folding up.

The amateurs of that time were the only ones making constructive advances in the industry. The commercial organizations were only interested in selling stock. The financiers used the radio companies as a means for hooking the public. The engineers had a hard time to build apparatus badly needed for the ships. I recall at one time when the United Wireless Company installed wireless equipment on many ships and the manager informed me that while the company sold millions of dollars worth of stock, he couldn't get enough money from the main office to buy transformers to fill orders. However, in order to fill these old orders, he gathered together all the old brass and junk in the plant which he sold to a junk man to get sufficient money to buy the necessary transformers. The sinking of the *S.S. Republic* and Jack Binns' heroic efforts to call for help, for the first time made the world radio conscious and it was thereafter that all the governments began to compel ships to install radio equipment, thus giving the companies a commercial opening.

(When the World War broke out I had an opportunity to exercise my abilities. I devised a means for locating submerged submarines and calculating their speed and direction, an idea that proved extremely valuable to France and Great Britain. Before the United States entered the war and before airplanes were even dreamed of as military weapons of any importance, I had demonstrated to my own government the practicability of wireless communication between airplanes in flight and the ground.)

De Forest finally moved to a little place on Sedgwick Ave., the Bronx, where with the assistance of Charles Gilbert he made little coils, crystals, and audions. The 3-stage amplifiers then weighed about 80 lbs. and sold for $500. When the war started, I purchased two of these 3-stage amplifiers from de Forest and presented them to the British and French governments. *They were the first 3-stage amplifiers seen in France* and the radio department of the French Navy was astonished to find amplification of 10,000 times in the 3-stage amplifier. These were built by de Forest in America.

Very few realized at that time the importance that the radio tube or so-called audion would play in the development of electronic science. It was the foresight of Mr. Folk of the American Telephone & Telegraph Co. who paid de Forest $250,000 for his radio tube inventions which enabled de Forest to do further work with this tube. It was no doubt the greatest contribution made to the A. T. & T. and a great deal of credit should be given to Mr. Folk in acquiring this valuable invention which later proved to be of inestimable value to the A. T. & T. I recall how everybody was surprised at the amount paid but it was the biggest bargain that A. T. & T. ever received.

From the angle of non-cooperation, cut-throat patent litigation, the industry has never grown up. It still hasn't learned how to cooperate as radio companies in various European countries have, so here we find a radio set selling for less than the cost of the tubes in any other country. The famous American institution of profitless business is part of the radio history.

JOHN L. REINARTZ—

Consultant, RCA Mfg. Co., Inc.

WHAT early amateur or constructor hasn't heard of the famous "Reinartz" circuit, or Reinartz's activities in furthering amateur radio! His leadership and fine work while with the American Radio Relay League *is recognized by many of its members, especially his efforts to help make the first amateur trans-Atlantic tests a success. A real old-timer and radio pioneer!*

FIFTY years since the start of radio. Thirty years since the writer, then a schoolboy, looked into the mysteries of wireless and found it interesting and fascinating. This fascination has persisted to this day. Many are the milestones that mark the passage of time. The first coherer, the first spark coil, the first earphone, the first radio tube, the first spanning of the Atlantic, the first short waves, the first ultra-short waves, and today experimental television. Who knows what the next 50 years will bring to the radio enthusiast? Surely anyone's guess is as good as another's. Radio's golden jubilee will bring many fond memories to the old-timers. It will also bring regret that the Father of Radio, Marchese Marconi, is absent.

PIERRE BOUCHERON—

Remington Arms Co.

ONE of the earliest amateurs in the New York area, Pierre Boucheron waged some of the first battles for radio's rights in the United States. He later became editor of Radio News *(then a Gernsback publication) and subsequently went to a high position with the Radio Corporation of America. Mr. Boucheron, though no longer active in radio, has earned a niche in its history.*

MY dubious claim to radio fame, if any, dates back 30 years when I was one of New York's pioneer wireless amateurs. Early in 1908, I came to the Big Town with a 1-inch spark coil purchased from the original E. I. Co., some dry cells, a regular telegraph key; I soon set up a 100-foot aerial atop a walk-up apartment on East 48th Street. With this outfit I proceeded to make life miserable for the night crews of the two nearby wireless stations who had to pound brass for a living, namely, the United Wireless, "NY" and the Brooklyn Navy Yard stations.

There was but a handful of local amateurs, of which there still live and thrive Major E. H. Armstrong of superhet fame, George Eltz, Ernest Amy, Louis Pacent, Joe Fried and several others whose present whereabouts and state of health are unknown to me. Oh yes, there was a millionaire playboy in the Bronx who owned, believe it or not, a 5 kw. etheric-power plant, and when he was "sending," every commercial, Naval and other operator on the entire Atlantic seaboard just stood by, groaned and waited for the din to cease and desist—which might be from an hour to a day. No rules, no regulations, no Government supervision. Just a lot of free-for-all, healthy fun, packed with a thrill a minute. Of course, we all knew who was who, and where. The professionals would telephone the boys (affluent ones who had 'em) and beg "please, old man, won't you *PLEASE* stand by for a few minutes; there's a ship

(*Continued on page* 624)

NEW TUBES INTRODUCED TO RADIO MEN —IN 1937-38

(Continued from page 621)

Of special interest are the low inter-electrode capacities which in carrier service constitute an element of unusual importance. It is also of interest to note that the heater voltages chosen for the types 310A and 311A tubes were based on the connection of 2 type 310A heaters in parallel and this combination operated in series with the heater of 1 type 311A tube. This arrangement results in high-efficiency operation in "regulated" offices where the battery voltage is held close to 24 V.

The characteristics data are available in Table X. Tube terminal connections are shown in Fig. 1.

(Data courtesy Bell Telephone Laboratories.)

329A. Except for its filament characteristics this tube is a counterpart of the type 311A. Filament rating: 0.85-A. at 7.5 V.

WL-706 Voltage Regulator Tube. There are 2 general types of regulator tubes, in one make, of which the WL-706 is representative; namely *voltage*-regulator tubes and *current*-regulator tubes. As both types are different in construction and principle as well as in application, they are described separately. Both types were illustrated in the article, "Keeping Voltage Constant", in August 1937 *Radio-Craft;* they are included in this later article for completeness of data-reference.

The WL-706 tube is a cold-cathode gaseous type. It, therefore, has a tendency to maintain its voltage drop at a constant value regardless of ordinary fluctuations in the supply line voltage. Thus it may be used to regulate the D.C. plate voltage applied to amplifier tubes; and in many experimental circuits. As the operation is entirely electronic the regulation action is practically instantaneous. Incorporated in the WL-706 is an auxiliary electrode which by special circuit connections initiates the ionization, thereby permitting the tube to break down and be used on lower voltage lines than would otherwise be the case.

Characteristics are given in Table XI. Terminal connections are shown in Fig. 1.

(Name of manufacturer will be supplied upon receipt of a stamped and self-addressed envelope.)

WL-896 Current-Regulator Tube. This tube is within its range a constant-current regulating device and differs from voltage regulators in that it contains a single electrode referred-to as the filament. This filament has a variable-resistance characteristic which tends to keep the current flowing through it constant regardless of the supply voltage.

The maintenance of a constant voltage, for instance on the filaments of radio or electronic tubes, is particularly

(Continued on page 625)

REMINISCENCES OF OLD-TIMERS

(Continued from page 622)

sinking off Hatteras and we're trying to get her position from 'HA' (Cape Hatteras)."

Wireless in those sweet, irresponsible days of our youth (I don't believe that the word "radio" was even coined then) was just a free, unbridled, bucking bronco, and wireless operators, commercial and Naval as well as amateurs, were a select crowd of swashbuckling young braggarts, of a do-or-die breed, comparable, I suppose, to the later-day crack-pots who flew Jennies with "OX-5" grinders over the French lines during the 'teens of the War. Ah me, those *were* the days!

Later, we New York amateurs became more numerous and troublesome, and as a consequence, there began to creep up some resentment from certain quarters about our noise-making on the air, so we formed a *protective* association headed by young W. E. D. Stokes, Jr., whose father was a power. We met and denounced this interference with our inalienable rights as free Americans. A clipping in my yellowing Scrap Book, dated November 30, 1910, reports an attendance of 20 ardent souls at the first and only meeting ever held to denounce the Depew bill "which proposes to require a Federal license for all wireless stations," and a resolution was passed to send young Stokes to Washington to see Congressman Whosit about it. Ah me, again!

Codes? Could we read, rapidly, what was going on in the air? Certainly. Some of us were land wire operators, and since the American Morse Code was then the more or less official wireless code of the U. S., it was easy to copy and gossip about everything that went on whether sacred or otherwise. What few foreign ships were equipped used the Continental Code. This was a cinch, too, for many of us had learned to read that Code as well as American Morse via the nightly broadcasts of "CC" (Cape Cod), a powerful Marconi station of the day that sent Marconigrams and press items to ships at sea, over and over again the whole night long. If you missed one word (speed was 12 w.p.m.) you got it on the next round.

And there was the now long-forgotten Navy Code, the first letter of the alphabet which I recall was 2 dashes, instead of a dot and a dash. Unlike the Continental and American Morse Codes which have many letters in common, there was no Navy character the same as any one of the other two codes. I suppose the master minds who originated this Navy Code sought extreme secrecy. They got it, too, for it isn't on record that any operator ever got beyond the speed of 5 words per minute with it, which of course soon meant complete oblivion for the Navy Code.

Finally, there came a day when it began to look as if the commercial and Naval authorities were going to have the last word about the right-of-way of the ether. Gloomy talk was heard about regulations, examinations and licenses, and of fines and even imprisonment for the willful boys bold enough to break the impending new laws. Then came that fateful International Conference of 1912. It looked bad for those unhampered, damped amateur waves. So those liberty-loving spirits among us amateurs of 1910-'12 who could not bear the thought of being muzzled sought Escape to the Tropics where we could pound away brass to our hearts' content amid the roar of the thunderous static that prevails there most of the year 'round.

We individually did this by casting aside drab land ties, such as school, job, or even sweetheart, and signing-on in the role of "Sparks" on board passenger, freight or tramp ships of the then scant American merchant marine. And what glorious years those were! Adventure, glamour, transient love—all mixed with the proper amount of storms, hurricane, wrecks, murderous captains, and other hazards of land and sea. Of course, I fell in easily with this motley crowd and there still hangs in my den Certificate of Skill No. 30, which was the very first attempt by the Government to license these sea-going radio rovers.

But that is another story, and since my old friend and one-time editorial chief, Hugo Gernsback, has confined me to a few hundred words, I leave you, oh, gentlest of readers, with a sigh for the past and a salute to the future of our beloved radio.

LOUIS GERARD PACENT—

Pres., Pacent Engineering Corp.

WHEN, more than 30 years ago, amateur radio was being born, one of the leading spirits in the new field was Louis Gerard Pacent, who had started his activities in 1906. He is a Fellow of the A.I.E.E., I.R.E., S.M.P.E., and Radio Club of America, where he held the office of President. Noted as a radio and sound engineer, he has contributed much to radio's progress.

IN the year 1865, Professor James C. Maxwell of the University of Edinburgh, Scotland, by a series of brilliant mathematical processes, attempted to show that such things as electromagnetic or radio waves must exist. Professor Maxwell predicted that electric oscillations in a circuit would produce electric waves in the ether and that these waves would travel with the velocity of light.

A few years later, Professor Heinrich Hertz, of Karlsruhe, performed experiments that proved definitely the existence of the radiations that were predicted by Maxwell's mathematics. Later, at the University of Bonn, Professor Hertz performed a number of classical experiments along these same lines. Most of these experiments were on wavelengths which even today are considered "short." He not only propagated these radio waves, but he detected them, yet he did not realize the tremendous possibilities of his achievements.

The man who had the vision necessary to see the possibilities in the then fairly well established theory of electric ether waves was Guglielmo Marconi. Marconi, at a very early age, began experimenting with radio apparatus and developing it along the lines which were to make it a commercial success. In June, 1896, he filed a patent with the Patent Office of Great Britain, covering "Improvements in Transmitting Electrical Impulses and Signals and an Apparatus Therefor." From this time on, Marconi was the leader in the new art, and due to the limitations of the apparatus so far developed it seemed to Marconi that the long wavelengths held the greatest promise for transmission over great distances. The other great minds of early radio concurred with Marconi in this belief, and the rapid commercial development of radio on the longer waves was a result. Accomplishment is also a strong argument, and perhaps for the reason that long-range communication on the longer wavelengths was already established, it was gradually assumed that such communication could not take place on the short wavelengths. As a matter of fact, this belief was so firmly established that when the Government assumed authority over the various agencies using the ether to transmit intelligence and allotted to the amateurs wavelengths of 200 meters and lower, it was at first thought to be a death blow to amateur radio. The amateurs, however, did not give up and soon proved that long-range communication on shorter and shorter wavelengths was not only possible, but possible with infinitely less power than that used on the long waves.

Many will recall the first signals on 200 meters, transmitted by radio station 1BCG at Greenwich, Conn., and received in England by Mr. P. F. Godley under the auspices of the Radio Club of America in 1920. (See *Proc., Radio Club of America*, Vol. 2, No. 2, February, 1922.) The idea of transmitting amateur signals across the Atlantic originated with the writer before the World War, when it was presented to the Board of Direction of the Radio Club of Amer-

(Continued on page 626)

NEW TUBES INTRODUCED TO RADIO MEN —IN 1937-38

(Continued from page 623)

important when these filaments are of the thoriated or oxide-coated type. This requisite constant filament voltage may be obtained either by connecting the WL-896 in series with the tube filaments as a ballast tube, or in series with the filament transformer primary, with secondaries delivering their requisite filament voltages. The tube also may be used to maintain constant voltage from storage batteries, particularly those in aircraft, or wind-electric or gasoline-generator service. Technical data are given in Table XII; tube terminals are shown in Fig. 1.

(Name of manufacturer will be supplied upon receipt of a stamped and self-addressed envelope.)

WL-788. This tube except for variations in its characteristics is a counterpart of the WL-896. Characteristics data are given in Table XIV. Tube terminals are the same as shown, in Fig. 1, for the WL-896.

CHARACTERISTICS DATA

XPD—TABLE I

Average Characteristics

Filament voltage	2 V.
Filament current	75 ma.
*Amplification factor	6.7
*Impedance	7,400 ohms
*Mutual conductance	0.9-ma./V.
Av. diode current	200 microamps.

*Taken at Ea=45 V.

XVS—TABLE II

Average Characteristics

Filament voltage	2 V.
Filament current	75 ma.
Amplification factor	400
Impedance	1.3 megs.
Mutual conductance (max.)	0.3-ma./V.
Mutual conductance (min.)	0.01-ma./V.

814—TABLE III

Tentative Characteristics and Ratings

Filament voltage (A.C. or D.C.)	10.0 V.
Filament current	3.25 A.
Transconductance, for plate cur. of 39 ma.	3,300 micromhos
Direct interelectrode capacities:	
Grid-Plate (with external shielding)	0.1-mmf. max.
Input	13.5 mmf. max.
Output	13.5 mmf. max.

Maximum Ratings and Typical Operating Conditions As R.F. Power Amplifier— Class B Telephony

Carrier conditions per tube for use with a max. modulation factor of 1.0

D.C. plate voltage	1,250 V. (max.)	
D.C. screen-grid voltage (grid No. 2)	300 V. (max.)	
D.C. plate current	60 ma. (max.)	
Plate input	75 W. (max.)	
Screen-grid input	6.7 W. (max.)	
Plate dissipation	50 W. (max.)	
Typical operation:		
D.C. plate voltage	1,000	1,250 V.
D.C. screen-grid voltage	200	200 V.
D.C. grid voltage (Grid No. 1)*	−28	−28 V.
Peak R.F. grid voltage	50	50 V.
Beam-forming-plate voltage†	0	0 V.

(Continued on page 627)

REMINISCENCES OF OLD-TIMERS

(Continued from page 624)

ica for consideration. The War prevented their carrying out these experiments at the time, but after the War, successful experiments were made and the results obtained astounded the radio world.

This brings to mind that in 1915 or thereabouts 100 to 200 meters were considered short waves and the general engineering opinion at that time was that these waves would not travel long distances. In recent years short waves of the order of 10 meters were also considered that they would not travel the curvature of the earth and only travel short distances. *Now both of these theories have been disproved and waves of 10 meters and less have covered half the World!* Everyone has now awakened to the possibilities of the short waves, and we now see radio once more attempting to return to the short wavelengths used by Hertz, so long ago.

IRVING VERMILYA, *W1ZE*—

Radio Station WNBH, *Standard-Times, Mercury*

RADIO "AMATEUR NO. 1" is Irving Vermilya, W1ZE, who started his experiments in 1901, and not so long afterward operated under the call letters, VN. In 1915 he received W1ZE as his call and has been using it ever since. Mr. Vermilya is also one of the first customers of the old E. I. Co., about the only place radio parts were obtainable in those early days of experimenting.

IT certainly gives me great pleasure to look back over the long years, and to recall some of the incidents and happenings of the years gone by. Certainly, a lot of "radio water" has passed under the bridge since I first became interested in radio in 1901 (and thus earned the sobriquet of "Amateur No. 1"—*Editor*). At that time I lived in Mt. Vernon, N. Y., and can well remember making trips to New York City in an effort to buy radio material, and I can also well remember that it was almost an impossibility to buy anything connected with radio in that day. Having been keenly interested in wire telegraphy, I naturally made the strongest effort to obtain wireless material at the J. H. Bunnell Co., which was located in Park Place, N. Y. C. As the years pass by, I have fond recollections of being able to buy radio equipment from Hugo Gernsback's Electro Importing Co.

I can also well remember that Mr. Gernsback launched probably what was then the first radio amateur call book in the world, and I was duly registered in this book under the call letters of "VN". This was many, many years before the Government even issued any licenses to any types of wireless stations and the wavelengths that we boys used in those days were anything from 200 to 900 meters. I recall with great pleasure having worked the following stations back in those dark dim years:

FK—Frank King; DR—Dr. Hudson; CC—C. Cannon; DX—Mr. Meyers; HB—A doctor in N. Y. C.; FH—Arthur Boeder; and WA—C. Runyon, Yonkers, N. Y.; and many others whose call letters I cannot now recall. Today under the call letters of W1ZE which I have now held for 22 years, I am still operating with a kilowatt on 5, 20, 80 and 160 meters—both phone and C.W.—and I would be very glad to contact any of my old friends.

Incidentally, some of the "newer-timers" may be interested to learn how I acquired the appellation of "Amateur No. 1." This was the title of an article I wrote, and which appeared some years ago in *QST*, in which I recounted how I had visited Marconi at Goat Hill and watched him fly his kites, and how, when I returned home, I then set up a coherer-decoherer and Ruhmkorff-coil amateur radio station—in 1901. I was 11 years of age. A year later I was presented with one of two coherers Mr. Marconi had given Dr. Tyndall.

JOHN V. L. HOGAN—

Pres., Interstate Broadcasting Co.

PROMINENT radio engineer and outstanding contributor to the progress of radio. Inventor of the gang-condenser (variable), President of the Institute of Radio Engineers in 1920 and owner and chief engineer of the renowned short-wave high-fidelity broadcast station WQXR. At the present time, Mr. Hogan is doing much valuable work in facsimile and high-fidelity transmission.

OF course, "radio" is a coined word, and it was not adopted until long after 1896, the year in which Marconi's work is widely considered to have given this new system of communication its real start. There is always some difficulty about deciding upon when "decades" begin and end; some people insist on having them in round figures, and so would put the First Radio Decade as from 1890 to 1900. I prefer to think of the First Radio Decade as starting with Marconi's work in 1896 and ending in 1906, and in looking back today, from the beginning of radio's Fifth Decade, I will classify it in that way.

America did not get into radio very effectively for several years after Marconi's first demonstrations abroad. I remember that in 1902 the wireless telegraph was not only a novelty, but a mystery. Perhaps my first real contact with the new science was when, in that year, I attended a showing of scientific marvels at the University of Indiana. Among these was a little spark-coil transmitter which was capable of operating a coherer and a relay across the length of one of the lecture halls. By pressing his telegraph key, the professor could make a sounder click, ring a bell, start a small electric motor, and so on. The climax of the evening was the setting off of a miniature cannon, *and the spark that ignited its gunpowder quite evidently also fired something within me, for I have been driven by "wireless" ever since.*

It was not long thereafter that I had the opportunity to spend many hours with the wireless operators at the U. S. Naval Station in San Juan, Porto Rico. This was equipped with a mercury-turbine interrupter spark-coil transmitter, and a coherer receiver, and, when the static was not too bad, exchanged messages with its sister station on Culebra Island, some 40 or 50 miles away. The chief operator, by name Jones, was more than patient under the barrage of questions that my youthful enthusiasm dictated, and I owe him a great debt of gratitude. Soon, then, as you may well imagine, I had my own and my first amateur wireless station. It was too feeble to communicate with neighbors, even had there been any who were interested in such impractical matters, but did effectively set up Morse Code buzzes in the telephone-and-microphone receiver that I had arranged about a hundred feet away.

Next came more realistic apparatus, including 60-cycle transformers, glass-plate condensers and tuning helices, together with more delicate receivers. In those days we amateurs were not regulated as to wavelength or power, and there was plenty of channel space in the radio spectrum. Interchange of messages was not among the amateurs alone; frequently one would find a commercial or Naval operator with whom he could "chew" during off-hours. George Davis at "PT" (the Brooklyn Navy Yard) and "Z" Duffy at "NY" (42 Broadway) had many amateur friends, and were ever willing to help us along with practical hints. Sometimes, however, one of the hams would set up too much "9", which was what QRM was called in the years before "Q" signals, and then we would all be unpopular for a time. There was no federal regulation, no broadcasting, not even any wireless telephone,—almost nothing but the bare frame of radio as we know it today.

(Continued on page 628)

NEW TUBES INTRODUCED TO RADIO MEN —IN 1937-38

(Continued from page 625)

*Grid voltages are given with respect to the mid-point of filament circuit operated on A.C. If D.C. is used, each stated value of grid voltage should be decreased by 7 volts and the circuit returns made to the negative end of the filament.

†Beam-forming plates should be connected to the mid-point of filament circuit operated on A.C., or to the negative end of the filament when a D.C. filament supply is used.

313A—TABLE IV

Ratings

Max. peak control-electrode current	30 ma.
Max. average control-electrode current (avg. over 1 sec.)	10 ma.
Max. peak reverse current in main gap	5 ma.

Characteristics

Nominal control-gap breakdown voltage	70 V.
Nominal control-gap sustaining voltage	60 V.
Nominal main-gap breakdown voltage	175 V.
Nominal main-gap sustaining voltage	75 V.
Transfer current	5 microam. (max.)
Nominal deionization time	
Main gap	10 millisec.
Control gap	3 millisec.

6J8G—TABLE V

Triode-Heptode Converter Characteristics

Heater voltage, A.C. or D.C.	6.3 V.
Heater current	0.3-A.

Operating Conditions and Characteristics

Heater voltage	6.3 V.
Plate voltage (heptode)	250 V. (max.)
Control-grid voltage (heptode)	–3 V.
Screen-grid voltage (heptode)	100 V. (max.)
Oscillator plate voltage (triode)*	250 V. (max.)
Oscillator grid resistor (triode)	50,000 ohms
Plate current (heptode)	1.2 ma.
Screen-grid current (heptode)	2.8 ma.
Oscillator plate current (triode)	5.0 ma.
Oscillator grid current (triode)	0.4-ma.
Plate resistance (heptode)	4.0 megohms approx.
Conversion conductance	290 micromhos
Control-grid voltage (heptode) for 2 micromhos conversion conductance	–20 V.

*Applied through 20,000-ohm dropping resistor.

6ZY5-G—TABLE VI

Tentative Data

Heater voltage	6.3 V.
Heater current	0.3-A

As Full-Wave Rectifier

Operating Conditions. Condenser Input to Filter

Heater voltage*	6.3 V.
A.C. plate voltage per plate (r.m.s.)	350 V. (max.)
Peak inverse voltage	1,000 V. (max.)
Peak plate current per plate	150 ma. (max.)
D.C. output current	35 ma. (max.)

When the D.C. resistance of each half of the secondary of the power transformer is less than 225 ohms, sufficient resistance should be added in series with each plate to provide this minimum of resistance.

*The D.C. potential between heater and cathode should be limited to 400 volts.

Under no condition of operation should the normal operating heater voltage of 6.3-volt ever fluctuate to exceed a maximum of 7.5 volts.

(Continued on page 629)

REMINISCENCES OF OLD-TIMERS

(*Continued from page* 626)

Just at the end of this First Decade, as I have placed it, I was happily given the position of laboratory assistant to Dr. Lee de Forest. He was then carrying forward his work on the audion, which up to that time had been principally a modified Fleming valve sensitized by the addition of a plate-circuit battery. *During my time with de Forest, the very first "grid audion"* (or 3-electrode vacuum tube) *was built;* the fact that I ran and took notes on the first plate-voltage—plate-current characteristic of this veritable but tiny grand-daddy of all radio amplifiers is perhaps one of the fondest of my early wireless memories. That particular little job was so significant, so prophetic of what afterward came to dominate radio, that I often wonder whether we are today doing anything that will be equally the basis of a great part of the radio practice that will exist 30 years from now. Perhaps we are; and if I were asked to put my finger on a single item of today's work that might contain the germ of an important radio service likely to flame into significance within the next 3 decades, I think that for that position of honor I would select *point-to-point*, and *broadcast* Facsimile.

J. R. POPPELE

Chief-Engineer, radio station WOR

TONS of expensive equipment, complicated apparatus, connected together by numerous heavy cables from studios to transmitters, to power plant room. Such is the picture of broadcast station WOR, of which Poppele is chief-engineer. It hurls 50,000 watts into the ether whenever the station is "on", and Poppele and his staff see to it that WOR functions smoothly and efficiently.

DESPITE the efficient, nation-wide industry which broadcasting has become today, there are few old-timers who cannot look back upon the early days before the war without a smile.

A personal recollection of my first encounter with radio-telephony is that it was a baffling experience. At the age of 13 or 14, like many of the other radio pioneers, I built an amateur wireless transmitter, operating under the call letters 2AEY at my home in Newark. It was a customary quenched-gap spark rig of the era, with a simple galena crystal detector as receiver—one of those patience-exacting outfits where one spent interminable minutes adjusting the all-important cat's-whisker to find the point of greatest sensitivity. My first wireless contact, I recall, was with another amateur across the Passaic River from my home.

Then one day in March, 1912, something new came into the headphones. Instead of the customary raucous spark, raspy as a file, I heard music—the strains of *Tipperary.*

Somehow, I felt this was impossible. So I searched the house for a phonograph, threw open the window and listened to see if the phenomenon were coming from a neighbor's house. But only in the headphones was the music audible.

A worried note to old *Modern Electrics* solved the mystery. There actually *was* a station accomplishing this seemingly impossible miracle of broadcasting music and voices by wireless and I had heard it.

These days of national networks and voices from the remotest cities of the globe are a far cry from March, 1912. And there are other recollections which crowd back upon us when we start reminiscing.

The wonderful Geissler tube, forerunner of the neon bulb, which I won for selling subscriptions to *Modern Electrics*, has not been forgotten. Nor have the halcyon days of the Great War when wireless altered the pattern and speeded the tempo of modern warfare. Then again, as commercial operator in the North Atlantic during 1919, I copied a message to watch for Alcock and Brown, the pioneer trans-Atlantic flyers. Imagine the thrill when I saw their frail plane pass us 10 miles off on the horizon and reported its progress to an anxious world.

And today, when such memories seem amusing, when such wonders have become commonplace, I am still proud to have been both a witness and a participant in the progress which has made radio a foremost factor in the web of modern communication and culture.

This concludes *Radio-Craft's* presentation of reminiscences of old-timers in radio. Although space limitations preclude printing the recollections of the many more pioneers who helped lay the foundation of modern radio, we salute you, one and all!

A $750, KC.-CALIBRATED TUNER—IN 1927!

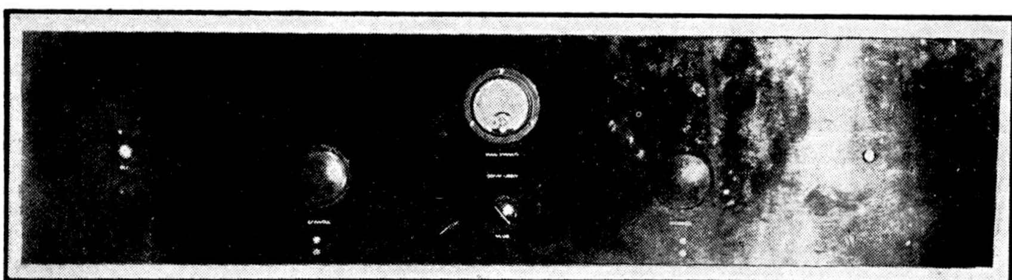

Front view. Note milliammeter for precision visual tuning!

Rear view. Printed-scale dials accurately calibrated in kc.!

A DECADE AGO, when custom set building was in its heydey, your Managing Editor—whose interest in radio is hereditary (Mr. Washburne, *pere*, having built his first spark coil and coherer-decoherer set in 1897)—joined Radio Construction Labs. and now presents the first published account of what was probably the finest engineering feat of that period. "I vas dere, Sharlie," and know whereof I speak.

Rudolph Siemens, in collaboration with Paul Laverty and business associates Howard Smith and John Murray, was producing a custom-built tuning unit that sold, sans audio system, for $750, and that could be *guaranteed* to pick up (even at West End Ave. locations in New York City) programs originating in California!

The final design of the tuner incorporated 5 stages of radio-frequency amplification (4 stages—as shown in photo at left—were used in early designs) in an improved Wheatstone-bridge version of the Rice neutrodyne circuit. A single, cast-metal dial, ***with accurate calibrations in kilocycles printed on the scale (!)***, was used (the earlier type—as shown in photo—used 2 dials). Each stage employed a type 201A tube, and was completely and individually shielded and bypassed; and could be individually cut into or out of circuit at the flick of a snap-switch. All 5 stages resulted in "California" sensitivity, and somewhat better than 10 kc. selectivity straight across the band; a lesser number of stages was used for local (less sensitive and selective) reception. Twin Hammarlund straight-line-frequency variable condensers were required in each stage; a tool-steel shaft turned all 10 condensers, without backlash, merely by blowing on the periphery of the tuning dial. A heavy, cast-metal base was found essential to maintain alignment.

National Union is making an extraordinary offer to all service men and dealers—a high quality tube tester, an oscillator and output meter, and a visameter. Three valuable pieces of radio service equipment . . .
All absolutely free. You can have any one or all of these pieces of equipment. Simply, easily and absolutely without charge.
This offer by National Union lasts for a short while only. Act at once!

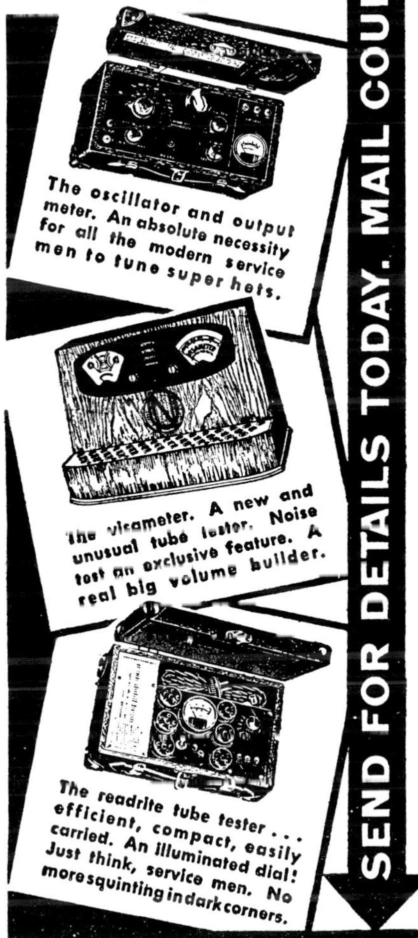

NATIONAL UNION RADIO CORP.
400 MADISON AVENUE, N. Y. C.
Please tell me how I can have any one or all of the above equipment free.
NAME________
ADDRESS________
CITY________ STATE________

REPRODUCED FROM OLD ADVERTISEMENT

NEW TUBES INTRODUCED TO RADIO MEN —IN 1937-38

(Continued from page 627)

884—Table VII

Tentative Data

Heater voltage†	6.3 V.
Heater current	0.6-A.
Direct Interelectrode Capacities:	
Grid to plate	3.5 mmf.
Grid to cathode	3.5 mmf.
Plate to cathode	2.5 mmf.
Tube voltage drop	16 V. approx.

As Sweep-Circuit Oscillator

Plate voltage (instantaneous)	300 V. (max.)
Peak voltage between any two electrodes	350 V. (max.)
Peak plate current	300 ma. (max.)
Average Plate Current:	
For frequencies below 200 cycles/sec.	3 ma. (max.)
For frequencies above 200 cycles/sec.	2 ma. (max.)

Grid Resistor—The resistance of the grid resistor should be not less than 1,000 ohms per maximum instantaneous volt applied to the grid. Resistance values in excess of 0.5-meg. may cause circuit instability.

As Grid-Controlled Rectifier

For frequencies below 75 cycles per sec.

Peak voltage between any two electrodes	300 ma. (max.)
Peak plate current	350 V. (max.)
Average plate current:	
Averaged over period of not more than 30 sec.	75 ma. (max.)

Grid Resistor—The resistance of the grid resistor should be not less than 1,000 ohms per maximum instantaneous volt applied to the grid. Resistance values in excess of 0.5-meg. may cause circuit instability.

Heater voltage should be applied for 30 seconds before drawing plate-load current.

†The cathode should preferably be connected directly to the mid-tap of the heater winding. In circuits where the cathode is not connected directly to the heater, the heater may be made negative with respect to the cathode by a potential difference not to exceed 100 volts provided the peak voltage between any electrode and the heater does not exceed 350 volts.

6G6G—Table VIII

Ratings

Heater voltage (A.C. or D.C.)	0.15-A.
Heater current	6.3 V.
Maximum plate voltage	180 V.
Maximum screen-grid voltage	180 V.

Amplifier—Class A

Plate voltage	135	180 V.
Screen-grid voltage	135	180 V.
Grid bias	-6*	-9* V.
Self-bias resistor	440	510 ohms
Amplification factor	360	400
Plate resistance	0.170	0.175-meg.
Transconductance	2,100	2,300 micromhos
Plate current	11.5	15 ma.
Screen-grid current	2	2.5 ma.
Load resistance	12,000	10,000 ohms
Total harmonic distortion	7.5	10 P.C.
Power output	0.6	1.1 W.

The voltage between heater and cathode should be kept as low as possible where they are not directly connected.

*Transformer or impedance input systems are recommended. If resistance coupling is used the D.C. resistance in the grid circuit should not exceed 0.5-meg. with self-bias. With fixed-bias the D.C. grid circuit resistance should not exceed 0.5-meg. under the 135 volt operating conditions or 0.05-meg. under the "180 volt" conditions.

(Continued on page 641)

GET READY TO CASH IN ON 1938 RADIO SERVICE BUSINESS

. . . the easy National Union way!

Get the equipment you need FREE with purchases of National Union radio tubes and condensers . . . .

WHAT do you need to make yours the "finest radio service lab in town"? Whatever it is, National Union is ready to give it to you. Thousands of service engineers are cashing in on the National Union free equipment plan. They're building their reputation . . . they're growing, making more money.
Now, with National Union condensers it's twice as easy as ever before because condenser purchases and N.U. tubes all help to earn the things you need. Get in on the N.U. proposition for 1938. Get in on it now! Ask for name of nearest N.U. distributor.

QUALITY! NATIONAL UNION'S BYWORD

So many thousands of Radio Service Engineers have proved National Union quality in the field that it is taken for granted. Due credit should be given to the large staff of research engineers who are constantly at work in the National Union Laboratories striving to improve N.U. Products and develop new and finer merchandise. The N.U. research staff is your assurance that National Union Products will never let you down.

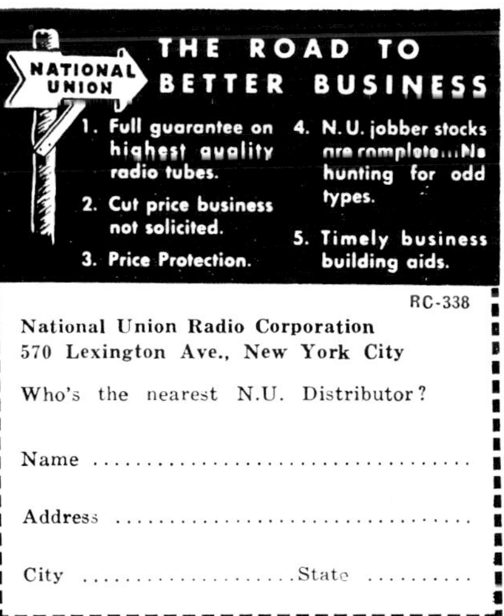

THE OLD E. I. Co. DAYS

(Continued from page 575)

orders fell off—as they have done every summer since that day. Coggeshall, my partner, thought that the wireless business was gone for good, and he offered to sell out to me. I accepted the offer and he departed for Chicago, leaving me as sole owner of the E. I. Co.

By that time we had moved to 87 Warren Street, a few blocks northwest from No. 32 Park Place where our first catalog had blossomed forth. The art work reproduced in its pages was done by Mr. Coggeshall, who was also an artist with a flair for beautiful women as exemplified on the cover of the first E. I. Co. catalog.

Several issues of this green-covered catalog were sent out thereafter, and soon trade was brisk again, and more and more people climbed up the 5 flights of stairs at 87 Warren Street to become initiated into the mysteries of *wireless.* Soon I brought out another catalog, on the cover of which I displayed the words "Modern Electrics"; and this, a few months later, became my first magazine.

"MODERN ELECTRICS"

Indeed, this was the first wireless or radio magazine in the world, and immediately started to make history. While the magazine was always operated separately from the E. I. Co., nevertheless the two were contemporary, and their fortunes were woven inextricably into each other. Neither the E. I. Co. nor the magazine made sufficient money to support me; but both of them together could support the organization, and gave the wireless amateurs of the day something that they could not have had otherwise. And it must be said in praise of those early amateurs that they responded nobly and supported my enterprises much better than could have been expected.

The service was not always of the best; the apparatus was still crude, and *Modern Electrics* in its first few issues had not achieved any great journalistic or literary efforts. But these were the pioneer radio days where anything went, as is usually the case when explorers sail uncharted waters. It soon became necessary to move from 87 Warren Street to 84 West Broadway, because the former quarters proved too small. It was really on the 4th floor of 84 West Broadway (which still remains today as it was in those days) that the E. I. Co. and *Modern Electrics* found their stride. The sinking of the *Republic,* with the heroic exploits of her wireless operator, Jack Binns, caught the imagination of the public, and a small boom for wireless immediately began. People were shouting for wireless, and we were there to give it to them. By that time it had improved somewhat, and we started to manufacture wireless apparatus in earnest.

In a single day I designed no less than 7 pieces of radio apparatus which became famous in their days. They comprised the Universal Crystal Detector Stand, the Double-Slide Electro Tuner, the Adjustable Transmitting Condenser, a sending Helix Inductance, the Electro Variable Sliding Plate Condenser, the Zinc Spark Gap, and the Electro Potentiometer, all of which are reproduced in these pages. These were all original designs and were, later on, copied by practically every wireless manufacturer of note. I still remember the day when I took my drawings over to a woodworking plant, and had them fashion the samples for all of the apparatus which made history later.

The only troublesome item was the potentiometer. Here I used the lead of a pencil as a resistor; and we used to buy pencils by the thousands, then soak them in hot water until the leads fell out. I originally had gotten hold of a number of Dixon pencils, and had a local laboratory test the resistance of various leads until I found one—a carpenter's flat pencil with a rectangular lead—of 300 ohms and another of 500 ohms. So we adopted these pencils and proudly announced the fact that, with each potentiometer, we furnished 2 rods, one of 300 and one of 500 ohms. We were blissfully ignorant of the fact that not all pencils would test alike; so that soon we began to get complaints that our 500 ohms were, maybe, 600 or 700, and our 300 ohms were either double that, or perhaps, only 50 ohms! How could we have known that pencil manufacturers were so careless in their "ohmage" when making pencils?

So I went to see the manager of the Dixon Crucible Company, makers of the pencils, and told him our requirements. Thereafter, they actually manufactured for us resistance rods which were accurate and dependable, so far as the ohmic resistance was concerned.

In leafing through one of the first issues of *Modern Electrics,* I came across an article written by me in the May 1907 issue of that magazine entitled "The Dynamophone." This now turns out to have been the first remote-control radio apparatus. As the article shows, I energized the transmitter by a microphone, whose impulses were superimposed on the spark and ensuing radio waves. At the receiving end, this operated the coherer and decoherer and an electric motor. It makes it thus the first remote-control wireless or radio apparatus on record. (See Fig. 45.)

RADIO "FIRSTS"

There were, however, other "firsts." For instance, in 1910, across the street at 69 West Broadway, I opened what was to become the first radio store the world had ever seen. This became the mecca of all wireless experimenters and amateurs of the day, and we really did a very huge business for the small compass of that store.

The only trouble was that we could not supply the material as fast as it was demanded. We had some rather famous customers at that store, too. Thus, we record, one of the sons of President Theodore Roosevelt, one of the grandsons of John D. Rockefeller, and many others.

I must not forget another important "first" and that is the vacuum tube or audion as it was called then, which was offered to the public through the E. I. Co. catalog in 1911. It was also advertised in the July 1911 issue of *Modern Electrics* by the E. I. Company. No vacuum tubes for radio purposes had previously been offered to the public.

The E. I. Company also sold parts for vacuum-tube amplifiers as early as 1913, mostly to jewelers all over the country. Many of them wrote in to us that with two audions hooked up with an E. I. Co. coupling impedance, they were able to pick up Arlington time signals (NAA) on 2,500 meters and amplify the signals with such volume that they could be heard a block away from the store. This was excellent advertising for the jewelers in those days.

It was also the E. I. Co. which sold probably one of the first, if not the very first, wireless telephone sets. This included a hand-fed French arc transmitter and was sold complete for $150 to a famous wireless experimenter by the name of Samuel Wein, who has since become well-known through his research work in photoelectric cells. Mr. Wein reported to us that he had successfully talked with this wireless telephone set at a distance of 4 miles, and later 7 miles, in Yonkers, New York.

In 1912-'14, arc transmitters for phone and code transmission, in the ratings of 1, 2 and 4 kilowatts, designed by H. W. Secor, were sold by the E. I. Company to a number of colleges and governments, including a 2-kw. arc which was bought by the Brazilian government.

There was another important "first": in the January 1909 issue of *Modern Electrics,* I announced the formation of the *Wireless Association of America.* It was actually the first national association anywhere, and it is believed that it was also the first wireless club in existence.

It is true that, in January, 1909, another radio organization was formed—the Junior Wireless Club of New York City. But inasmuch as the January issue of the magazine came out in December, and the arrangements for the Association had been made in November and December of 1908, it is most likely that the honors of "first" should go to the *Wireless Association of America.*

In any event, it was the first *national* organization to bring amateurs from all over the country together under one head. Let me quote from that excellent book "Two Hundred Meters and Down, the Story of Amateur Radio," by Clinton B. De Soto, Assistant Secretary of the American Radio Relay League:

"*The Wireless Association of America* was a child of Hugo Gernsback, publisher of *Modern Electrics.* After the first few months of its existence, Gernsback announced a membership totalling 3,200. By November 1910, he claimed

A Wonderful "NEW VOICE" for Radio

The New QUAM CONE SPEAKER

Retails at $17.50

The Quam Loud Speaker introduces a new and startling improvement in Loud Speaker construction. Instead of a floating reed secured at only one end, the Quam Speaker has a *Stretched Reed*—like the human vocal cords. The result is amazingly superior to any you have heard before.

Compare This Speaker with Any Other

Selling at TWICE its Price!

QUAM RADIO CORPORATION

9718 Cottage Grove Ave.
Chicago, Ill., U. S. A.

THIS ADVERTISEMENT APPEARED IN 1927

that this number had jumped to 10,000. It was easy to recruit members for such an organization; there were no dues and no obligations. Youthful electrical experimenters signed up in swarms, attracted by the famous names in the honorary membership group and the ease of becoming a member. The membership represented a fairly accurate index of national interest in radio, although not, of course, of the number of active transmitters. Even so, the number of worthwhile amateur stations on the air had, according to conservative observers, increased from perhaps one hundred and fifty in 1905 to five or six hundred. The number of small spark coils in use was several times that figure.

"In early 1910 the first *Wireless Blue Book* of the *Association* appeared, dated 1909. It listed ninety U. S. amateur stations who were members of the *Association*, together with the call letters used, approximate wavelengths in meters, and the spark length of the induction coil. Stations were listed in the following states: Massachusetts, N. Y., N. J., No. Carolina, Mo., Calif., Texas, R. I., Oregon, Ill., Ohio, Pa., Conn., Fla., Ind., West Va., Montana, Wash., Minnesota, Wisconsin, and Maine. Wavelengths ranged from 32 to 950 meters. The average spark gaps were from ½ to 3 inches. Two stations had the exceptional length of 10 and 14 inches, respectively.

"The second *Blue Book* appeared June 1, 1910. Meanwhile, the number of copies of *Modern Electrics* printed had increased from the initial 2,000 to 30,000. The *Wireless Association of America* continued to send out more and more gaudy membership certificates, and the cumulative members on the membership rolls mounted higher and higher."

Directors of the Wireless Association of America were as follows:

Dr. Lee de Forest, President
John Stone Stone, Vice-President
Wm. Maver, Jr., Secretary
Hugo Gernsback,
Chairman and Business Manager

In the January 1909 issue of *Modern Electrics* I said, editorially:

"It affords the Editor genuine pleasure to offer to the thousands of wireless enthusiasts an appropriate New Year's present in the form of the *Wireless Association of America*. Every reader perusing the announcement of the *Association*, found elsewhere in this issue, will agree that such an organization has been needed for a long time.

"Wireless Telegraphy may now be said to have left the experimental stage. It has become an everyday necessity, and already rivals the wire and cable telegraph. Between sea and land, wireless is of the utmost importance and today forms as necessary an adjunct to every up to date vessel, as its coal and lifeboats.

"Only a few days ago Wireless demonstrated anew its utility. Several Italian warships were at sea, when the disaster occurred in Southern Italy, necessitating prompt assistance by outside help. The warships were at once recalled by means of wireless, and through it, saved the lives of hundreds of sufferers and brought the much needed relief.

"Not alone on sea, or between land and sea, but on the mainland also, Wireless is of high importance. Only a few weeks ago two young Wireless experimenters saved several hundred lives in southern France in a dramatic manner. One of the young men, living on a hill near a dam, discovered that it was giving away, which meant the destruction of a small village in the valley below and the drowning of hundreds. He promptly "called up" his friend, living in the village below, who gave the alarm. Over two hundred people reached the mountain in safety, just in time to avoid the flood. Without Wireless they would have been surely drowned, as telephone and telegraph are unknown in that locality and the person in charge of the dam was sick on account of a recent accident."

In announcing the *Wireless Association of America* in that issue of *Modern Electrics*, I said:

"The *Wireless Association of America* has been

(Continued on following page)

THE OLD E. I. Co. DAYS

(Continued from preceding page)

founded with the sole object of furthering the interests of wireless telegraphy and aerophony in America.

"We are now on the threshold of the wireless era, and just beginning to rub our intellectual eyes, as it were. Sometimes we look over the wall of our barred knowledge in amazement, wondering what lies beyond the wall, as yet covered with a dense haze.

"However, young America, up to the occasion, is wide awake as usual.

"Even Dr. Lee de Forest, America's foremost wireless authority, confessed himself surprised that so many young men in this country should be in the possession of such well-constructed and well-managed wireless stations; which is only another proof that the clear-headed young men of this country are unusually advanced in the youngest branch of electrical science.

"So far America has led in the race. The next thing is to stay in the front, and let the others follow. In fact he would be a bold prophet who would even dare hint at the wonders to come during the next decade.

"The boy experimenting in an attic today may be an authority tomorrow. However, not even the cleverest inventors or experimenters always have the opportunity of making themselves known to the world, and it is right here that we are confronted with a mystery so far unsolved.

"As stated before, the new *Wireless Association's* sole aim is to further the interests of experimental wireless telegraphy and aerophony in this country.

"Headed by America's foremost wireless men, it is not a money-making institution. There are no membership fees, and no contributions required to become a member.

"There are two conditions only. Each member of the *Association* must be an American citizen and must own a wireless station, either for sending or for receiving, or both.

"The Board of Directors of this *Association* earnestly request every wireless experimenter and owner of a station to apply for membership in the *Association* by submitting his name, address, location, instruments used, etc., to the business manager. There is no charge or fee whatever connected with this.

"Each member will be recorded and all members will be classified by town and state."

The first emblem of the *Wireless Association*, a button, is also reproduced here for the sake of record.

About this time, and even previous to that, I had started to print a number of radio handbooks, which were the first of their type to appear in print. Notable among them were:

"How to Make Wireless Instruments (Symposium)"; "The Wireless Telephone," by H. Gernsback; "Construction of Induction Coils and Transformers," by H. Winfield Secor; "Wireless Hookups," by Hugo Gernsback and G. E. Rudolph.

These were all 80- to 96-page radio handbooks.

"The Wireless Telephone," the first book on radio telephony to appear anywhere in print, was published by me in February, 1910, and contains a rather astonishing preface, which I reprint here, merely to show the trend of thought at that time:

"The present little volume is intended for the experimenter doing research work in wireless telephony and the student who wishes to keep abreast with the youngest branch of the wireless art.

"The author realizes that the future use of the wireless telephone will be confined to the lower power or battery system; as the present instruments, necessitating 220 and 550 volts for their successful operation, are not desirable nor practical enough for everyday use.

"The wireless telephone of the future must be as flexible as the wire telephone of today.

"Every farmer will be able to operate his wireless telephone, when the sending and receiving instruments will be housed in a box a foot square, without depending on the lighting current for its operation.

"The author predicts that, in less than 10 years, this stage will have been reached; as it is bound to come sooner or later."

Remember, in those days we had only spark coils and crystal detectors, the vacuum tube as yet being an unknown factor—Lee de Forest just having constructed his first audion which, by the way, is illustrated on page 69 of the book mentioned above.

But the radio telephone—which was to become broadcasting later on—had not as yet been invented. Radio, or rather wireless, in those days was inarticulate. We could only send and receive dots and dashes; the transmission of the human voice was still in the future.

We dimly suspected that, sometime or other, it would come about. In the meanwhile, we valiantly experimented with our spark coils, and often substituted a microphone for the key, only to hear hashed-up voices or hashed sounds over our electrolytic detectors (or crystal detectors) and our earphones.

Once in a while, if we were lucky, we could hear part of a word, but usually it was broken up badly. I knew what the requirements of the final radio telephone were; and the diagram which I reproduce herewith from page 13 of my book, "The Wireless Telephone," shows a capital "L" with a question mark in a box. This, then, was the "missing link." The question mark later became the vacuum tube and, if this is now inserted in the circuit at "L", it will be seen that this fits in closely with the elementary requirements of our present-day radio telephone.

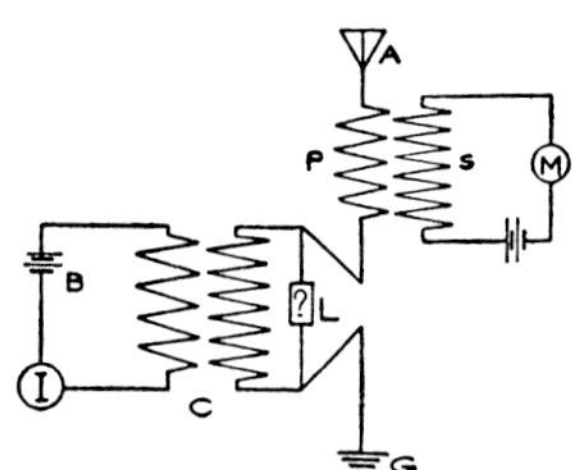

About 1908, I had begun to expand my activities by trying to introduce wireless instruments to the rest of the country. I started traveling a good deal, and began to introduce my wireless instruments and apparatus to the different radio stores as far west as St. Louis.

Naturally, this new amateur wireless industry attracted many boys of talent, and indeed it has been said that I have employed more people who later became famous in the radio industry, than perhaps any other man. This may or may not be true; but here is a partial list of those who worked with me in the old E. I. Company days, and who have helped to make radio what it is today:

EMPLOYEES OF YESTERYEAR

There was, first of all, Austin C. Lescarboura, who became the first managing editor of *Modern Electrics*, and later managing editor of a number of other publications.

Victor H. Laughter was another importation of mine. I brought him all the way from Mississippi after I had seen some of his letters. He was an excellent radio man and is today a manufacturer of radio testing equipment.

Isidor Goldberg, like Lescarboura, came to me in knee pants, and was active in our factory in various capacities. This early radio training never left him, and he is, today, the president of the Pilot Radio Corporation.

About this time also, Harry W. Secor came with me in an editorial capacity; this veteran radio man has ever since been with me, and is today managing editor of *Short Wave & Television*.

A little later, E. R. Weedon became the general manager of the E. I. Company; he is today advertising manager of Butterick Publications. It was he also who, in 1911, induced me to put out the E. I. Company's first Spanish catalog, as he had been in South America and was well acquainted with the Spanish language. Over 100,000 catalogs in Spanish were circulated in South America thereafter, and the venture proved profitable. As far as is known, this was the first radio catalog in Spanish.

Then there was Milton B. Sleeper, who first came with me in the capacity of associate editor of the *Electrical Experimenter*.

This well-known radio engineer is now domestic sales manager for F. A. D. Andrea, radio receiver manufacturers.

Next on the list is H. G. Cisin, who started with us and is well-known as a technical radio writer. Incidentally, it was he who invented the first A.C.-D.C. radio set, and his patent was recently upheld.

From Down East I imported a 100% Yankee of the taciturn type of Calvin Coolidge. His name is Clyde Fitch, who was with us for a number of years in various capacities, in the radio laboratory designing radio circuits and radio instruments, and also associate editor of the *Experimenter* magazine, as well as *Everyday Mechanics*. He became somewhat famous for his superheterodyne circuits, especially the Tropadyne, which was in high vogue during the late 'twenties.

Then there was Robert Hertzberg, who was on the staff of several of my magazines including *Television News* (later merged with my present *Short Wave & Television* magazine). Mr. Hertzberg is now editor of *Modern Mechanix* magazine.

Later on, through the E. I. Co. or through my various magazines, other names which since have become famous were added to the list of those who started with me. Among them are such well-known radio names, as Samuel

RADIO AT NEW YORK PRICES

Via Daily Parcel Post

The Prices Quoted Below Deliver Goods to Your Door

REFERENCES:
R. G. Duns, Bradstreet, Coal & Iron National Bank, Corn Exchange Bank, N. Y. City. No salvaged goods handled. Only standard brands in original packing, which bears ours and the manufacturers' guarantee. Due to our tremendous output we can undersell any of our competitors. Send your order in today and insure prompt delivery.

HOW TO ORDER—Write your order plainly; state number, description and price of items wanted. Send Post Office or Express Money Order, personal check or bank draft.

VACUUM TUBES

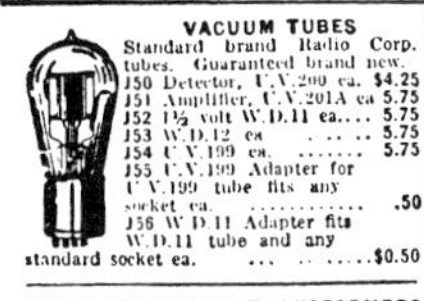

Standard brand Radio Corp. tubes. Guaranteed brand new.
J50 Detector, U.V.200 ea. $4.25
J51 Amplifier, U.V.201A ea 5.75
J52 1½ volt W.D.11 ea.... 5.75
J53 W.D.12 ea 5.75
J54 U.V.199 ea. 5.75
J55 U.V.199 Adapter for U.V.199 tube fits any socket ea. .... .50
J56 W.D.11 Adapter fits W.D.11 tube and any standard socket ea. $0.50

AUDIO FREQUENCY TRANSFORMERS

The following transformers are guaranteed standard makes, and will produce very efficient results
J63 Dietzen 3 to 1 Ratio. $3.95
J64 Dietzen 5 to 1 Ratio 3.95
J65 Dietzen 10 to 1 Ratio 3.95
J66 Acme Audio 4.25
J67 Amertran 5.95
J68 U.V. 712 R.C.A. 5.75
J69 U.V. 714 R.C.A. 5.75

LOOP AERIAL

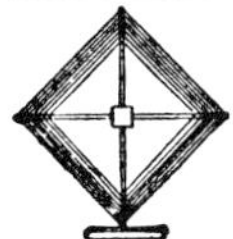

J76 Can be assembled by anyone in five minutes, all wood parts, wire and binding posts included, complete $1.00

MOULDED BAKELITE VARIOMETER

J79 Ea. $5.75. List...$8.00
A high grade instrument. Takes in the wave lengths from 250 meters to 800 meters. Table or Panel Mounting.

FRESHMAN VARIABLE GRID LEAK AND GRID CONDENSER

Lowest filament current increases battery life; eliminate howling; zero to 5 megohms.
J102 Freshman Variable Grid Leak alone.......... .....$0.65
J103 Freshman Variable Grid Leak and Condenser combined .... .85

FILAMENT CONTROLLED RHEOSTATS

Highest grade of material used, tapered knob.
J104—6 ohm .$0.39
J105—30 ohm ...... .49
J106—200 ohm Potentiometer .... .79
J107—400 ohm Potentiometer .79

WAVE TRAP

Simplifies tuning. Eliminates interfering stations, thereby improving the selectivity of your set. Selects between conflicting stations. Special while quantity lasts.
J131 Each $4.95

HONEYCOMB COIL MOUNTINGS

J140 2-coil mounting $2.85
J141 3-coil mounting 3.79
J142 Single coil mounting ...... .40
J143 Receptical for single coil mounting .... .50

MISCELLANEOUS

J163 Spaghetti Tubing per yd......$0.09
J164 Synthetic Crystal .... .30
J165 Ground Clamps .... .17
J166 Black Rubber Binding Posts .. .05
J167 Nickel Plated Binding Posts 2 for .05
J168 No. 18 Annunciated Wire half pound coil .... .39
J169 6 ft. Phone Cord with tips..... .79
J170 20 ft. Extension Cord with tips 1.95

HARD RUBBER PANELS

The highest grade panel on the market. Highly polished finish; cut in the following sizes:
J88—7x10 $0.95
J89—7x12 ... 1.15
J90—7x14 ... 1.25
J91—7x18 ... 1.65
J92—7x21 ... 1.95
J93—7x24 ... 2.20

HIGH GRADE MOULDED VARIOMETERS

J97 Price $2.65
A real high grade job at extremely low price. Takes in all wave lengths.
J98 Wooden Variometer same as above type $2.65

IMPROVED OUTDOOR LIGHTNING ARRESTER

Passed by the Board of Fire Underwriters. Regularly sells for $1.50.
J108 ....$0.95

AMPLITONE LOUD SPEAKERS

This speaker will give efficient results when used with any pair of Head Phones; highly nickel plated; 18 inches high.
J117 ....$3.95

STANDARD BRAND HEAD PHONES

J134 Brandes Superior $5.25
J135 Dictograph 6.75
J136 Baldwin Type "C" Double .. 9.85
J137 Baldwin Type "C" Single .. 4.95
J138 Western Electric 10.50
J139 Dietzen 3000 ohms extra special...... 3.95

WOOD CABINETS

Highest grade Mahogany finish Cabinets Hinged top. These cabinets are being sold at less than one-half regular value.
J57 Wood Cabinet; panel size 7"x10" Each $2.50
J58 Wood Cabinet; panel size 7"x12" Each 2.75
J59 Wood Cabinet; panel size 7"x14" Each 3.00
J60 Wood Cabinet; panel size 7"x18" Each 3.50
J61 Wood Cabinet; panel size 7"x21" Each 3.75
J62 Wood Cabinet; panel size 7"x24" Each 4.00

RADIO FREQUENCY TRANSFORMERS

J70 Tri-coil for 201A or 301A tubes $1.95
J71 For 199, 299 or W.D.11 and 12 tubes 1.95
J72 Erla AB1 .... 3.75
J73 Erla AB2 ....3.75
J74 Erla AB3 .... 3.75
J75 Erla Reflex 4.65

VARIOCOUPLER

J77 High Grade 180 Degree Coupler made of Bakelite tubing wound with green silk wire. Special $2.65
J78 Variometer same specifications $2.65

PLATE CIRCUIT "B" BATTERIES

The highest grade of battery made at prices lower than they ever were sold before
J80—22½ Volt Small $1.15
J81—22½ Volt Medium 1.45
J82—22½ Volt Large.. .... 1.95
J83—45 Volt Medium 1.95
J84—45 Volt Large 2.50
J85—45 Volt Extra Large.......... 3.50
J86—360 hr. 1½ Volt twin double duty "A" Battery .... .75

RUSONITE CRYSTAL RECTIFIER MULTIPOINT

J159 The perfect synthetic crystal detector sensitive over entire surface. Loud and clear. Price, mounted 50c
J160 14 K. Gold Supersensitive Rusonite Catwhisker. Permanent. Will not oxidize. Price 25c
J161 Rusonite Reflex Crystal. Manufactured expressly for Reflex Circuits. Guaranteed. Price, mounted $1.00

JACKS AND PLUGS

Jacks are polished nickel constructed with pure silver contacts.
J118 One Spring open circuit... ...$0.39
J119 Two spring closed circuit...... .49
J120 Round plug as illustrated..... .49

LOUD SPEAKERS

J125 Herald Loud Speaker; no extra battery required. The ideal instrument for the home, school or church. New .$30.00
J126 Magnavox new type $29.95
J127 Music Master 28.50
J128 Atlas Speaker 22.50
J129 Pathe Speaker 21.95
J130 Dietzen Speaker 18.95
The above prices delivered to your door.

1½ VOLT DIETZEN TUBE

Two-element Fleming valve effect; operates on one dry cell with or without "B" Battery; wiring diagram free with each tube.
J96 Socket for Dietzen Tube .... .40

BAKELITE TUBE SOCKETS

Moulded of brown bakelite, Binding Post Connections.
J99 Standard Type for 201A 301A, W.D.12 .... .39c
J100 For U.V.199, U.V.299 Socket .... .39c
J101 For W.D.11 .. .39c

RADIO SOLDERING IRON

Indispensible for the man who builds his own sets.
J109 ....$2.50
J133 Soldering Paste Tube .25

VARIABLE AIR CONDENSERS

These condensers are made of heavy aluminum plates, high grade bakelite ends.
J110—21 Plate .0005 Mfd. ..$1.75
J111—11 Plate .00025 Mfd. .. 1.25
J112—3 Plate .00005 Mfd. $0.95
J113—43 Plate .001 Mfd. 2.25

VERNIER VARIABLE CONDENSERS

Specifications the same as above.
J114—11 Plate Vernier with knob and dial $2.95
J115—26 Plate Vernier with knob and dial 3.45
J116—43 Plate Vernier with knob and dial 3.95

MAGNET WIRE (8 oz.)

Double Cotton Covered	Size	Price	Green Silk	Size	Price
J144	18	$0.43	J145	18	$0.73
J146	20	.53	J147	20	.85
J148	22	.68	J149	22	1.00
J150	24	.78	J151	24	1.13
J152	26	.88	J153	26	1.65
J154	28	1.07	J155	28	1.95
J156	30	1.58	J157	30	2.65

Modell's 58 Cortlandt St. New York, N. Y.
Since 1889
America's Greatest Radio Mail Order House
ELEVEN NEW YORK STORES

THIS ADVERTISEMENT APPEARED IN 1923

Cohen, now president of the General Instrument Corporation. He still manufactures enormous quantities of radio condensers, for practically every radio set made in the United States today.

W. G. Maney, now with Aerovox Company;

A. P. Peck, now associate editor of *Scientific American;*

Maurice L. Muhleman, editor of a number of radio magazines, and now of *All Wave Radio;*

Leon L. Adelman, now sales manager of Cornell-Dubilier;

John F. Rider, now publisher of "Rider's Radio Manuals";

Alfred P. Morgan, who later on became a radio manufacturer in his own right;

E. J. Quinby, as well as C. W. Palmer, both of whom are now with the Western Electric Company;

Then there was E. T. Jones, one of my first editors of *Radio News,* now in the manufacturing division of RCA at Camden, N. J.;

Pierre Boucheron, also one of my first editors on *Radio News,* who later became well-known as an official of the RCA Radio Mfg. Corp., and now general merchandising manager of Remington Arms Manufacturing Co.

And I must not forget Robert E. Lacault, who came to me, fresh from France, after the war, to become one of my best editors on *Radio News.* Mr. Lacault, inventor of the Ultradyne circuit, was one of the outstanding designers of superheterodynes. He later on became a manufacturer of radio sets, and a brilliant future was predicted for him, but an untimely death cut his career short.

There were many others who, if they did not become so prominent, are still connected with the radio industry in one way or another.

Before the World War, the E. I. Company became known through every part of the world, and there are few countries today which can not boast of some early E. I. Co. apparatus. Many of these articles, by the way, were patented by me and, at the height of the E. I. Co.'s career, I had a string of patents, all pertaining to wireless apparatus of one sort or another.

One of the best known patents, no doubt, was that of the Gernsback Slider, used in all E. I. Co. tuning coils. This was made of red and black composition and was made to slide on a square rod; the contact was effected by a ball, attached by a spring to the slider.

Another familiar patent of the E. I. Co. was an electrolytic interrupter, which consumed a terrific amount of current and merrily burned out fuses all over the land.

Then there was the well-known Electro Rotary Potentiometer, also patented by me. Also, I had a compression chuck headband with which most of the later E. I. Co. headsets were equipped. In addition to this, minor other patents helped to give the E. I. Co. its reputation.

There was also the famous Rheostat

(Continued on following page)

THE OLD E. I. Co. DAYS

(Continued from preceding page)

Regulator, which was built with an open coilspring resistance for better cooling. Almost every experimenter in the country had one of these rheostats.

I must not forget another E. I. Co. piece of apparatus which became famous—the *Radioson* sealed-in electrolytic detector, also reproduced in this article. This had been developed by me, and it certainly was (and still is, to my mind) the most sensitive detector except for the vacuum tube.

I developed a method whereby a special type of Wollaston wire was fused into a glass tube and then drawn out into a sealed tube. This was then broken in half, and the sealed half ground on a very fine stone, and then highly polished. Each point was then viewed, through a high-power microscope, by several operators before the detector part proper was sealed in a glass tube containing diluted sulphuric acid.

While these detectors were satisfactory, the trouble was that strong static tended to burn them out inasmuch as the fine platinum Wollaston wire was less than 1/10,000 of an inch in diameter.

One of the things which helped to disseminate wireless in the early days was the E. I. Co.'s famous "Wireless Course in 20 Lessons," by S. Gernsback, H. W. Secor and A. C. Lescarboura. This "Wireless Course" went through more than 20 editions, totaling some 200,000 copies, and was given away free through the E. I. Co. catalog in a rather novel manner. Bound in the middle of the catalog were 20 coupons; if you needed wireless supplies, you tore out one of the coupons which entitled you to one lesson, provided your order was for $1.00 or more; if you ordered merchandising totaling $20, you, therefore, received the 20 lessons free.

The E. I. Co. proclaimed proudly that the "Wireless Course in 20 Lessons" was "In use on every battleship of the U. S. Navy." Indeed, the course was used to teach newly recruited Army and Navy operators during the World War.

RADIO ("WIRELESS") LEGISLATION

The part played by me in the enactment of wireless legislation has been printed many times by others. I give herewith a reprint from one of the later catalogs of the E. I. Co., which states the case concisely. It must be remembered that there were then no radio laws, nor restrictions, and any one could set up a transmitter and raise a rumpus in the so-called "ether," to the annoyance of amateurs and government stations as well. This state of affairs could not, of course, go on. I clearly foresaw that, sooner or later, the government would put a halt to all amateur radio activities, and in *Modern Electrics* I sounded many warnings to this effect. The catalog reprint follows:

The very first talk about Wireless Legislation in the country started in 1908. The writer in his Editorial in the November, 1908, issue of **Modern Electrics** pointed out that a wireless law

WIRING DIAGRAMS

package number 1 supplement to John F. Rider's **"Trouble Shooter's Manual"** is now ready for distribution. Here is an opportunity to secure wiring diagrams of the modern screen grid receivers at a ridiculously low price. 115 wiring diagrams, size 8½ in. x 11 in. punched three holes suitable for loose leaf binding, covering screen grid receivers produced by such manufacturers as Grebe, Crosley. Stromberg-Carlson, Fada, Stewart-Warner, Edison, Eveready, Silver-Marshall and others . . . Price $2.50 postpaid. . . . Write for list of diagrams and books.

RADIO TREATISE COMPANY

1440 BROADWAY NEW YORK CITY

REPRODUCED FROM OLD ADVERTISEMENT

was sure to be passed in a very short while. In order to guard against unfair legislation as far as the wireless amateur was concerned the writer, in January, 1909, organized the "Wireless Association of America." This was done to bring all wireless amateurs together and to protest against unfair laws. Previous to this time there was no wireless club or association in the country. In January, 1913, there were over 230 clubs in existence, all of which owe their origin to the "Wireless Association of America."

The association had no sooner become a national body than the first wireless bill made its appearance. It was the famous Roberts Bill, put up by the since defunct wireless "trust." The writer, single-handedly, fought this bill, tooth and nail. He had representatives in Washington, and was the direct cause of having some 8,000 wireless amateurs send protesting letters and telegrams to their congressmen in Washington. The writer's Editorial which inspired the thousands of amateurs, appeared in the January, 1910, issue of **Modern Electrics. It was the only Editorial during this time that fought the Roberts Bill.** No other electrical periodical seemed to care a whoop whether the amateur should be muzzled or not. If the Roberts Bill had become a law there would be no wireless amateurs today.

That Editorial quickly found its way into the press and hundreds of newspapers endorsed the writer's stand. During January, 1910, the **New York American**, the **New York Independent**, the **New York World**, the **New York Times**, the **Boston Transcript**, etc., all lauded and commended the writer's views. (See Editorial article February, 1910, **Modern Electrics.**) Public sentiment quickly turned against the Roberts Bill and it was dropped.

Later, the Alexander Bill made its appearance, on December 11, 1911. This bill as far as the amateur was concerned was not quite acceptable to the writer, who had the amateurs' rights at heart, and steps were immediately taken to bring about an amendment as the writer, perhaps more than anyone else, realized that this bill, in some form or other, would become a law sooner or later. This is clearly stated in his Editorial in the February, 1912, issue of **Modern Electrics. In that Editorial is to be found also the first and now historical recommendation that if a wireless law was to be framed it should restrict the amateur from using a higher power than 1 kw. and his wavelength should be kept below 200 meters.** No one else had thought of it before and it is to be noted that when Congress finally passed the present wireless law, **it accepted the writer's recommendation in full,** thus paying him the greatest compliment, while at the same time acknowledging the fact that he acted as the then **sole** spokesman for and in behalf of the wireless amateur.

In March, 1912, the writer in a letter to the **New York Times** (See page 24, April, 1912, issue **Modern Electrics.**) pointed out the shortcomings of the Alexander Bill, and protested against unfair legislation.

The Times, as well as a host of other newspapers, took up the cry and published broadcast the shortcomings of the Alexander Bill. All this agitation had the desired effect and Mr. Alexander for the first time realized that the amateur could not be muzzled, especially when there was such a periodical as **Modern Electrics** to champion his cause. Promptly in April the Alexander Wireless Bill, **amended**, appeared and here for the first time in history the amateur and his rights are introduced in any wireless bill.

Mr. Alexander and his advisers accepted the writer's recommendation as set forth in his Editorial in the February, 1912, issue of **Modern Electrics.**

It will be noted that it copied the writer's recommendations word for word. The amateur had at last come into his own. This is all the more remarkable as this is the only country that recognizes the wireless amateur. On May 7, 1912, the Alexander Bill, amended, now known as S-6412, passed the United States Senate and on May 8th was sent to the House of Representatives and referred to the Committee on the Merchant Marine and Fisheries. The bill was signed on August 13th by President Taft, thus making it a law.

This terminated the fight which the writer had waged single-handedly for almost five years in behalf of the American amateur.

With wireless having become more specialized under government regu-

(Continued on following page)

THE OLD E. I. Co. DAYS

(Continued from preceding page)

lations, radio began to expand rapidly and, while the E. I. Co. had its ups and downs, still it continued to make progress. Came the World War and, with the entrance of the U. S. into the war in 1917, the official edict that all radio activities must cease immediately. No further radio instruments could be sold; so the puzzle for the E. I. Co. became "What to do?"

For the time being, it was necessary to change the entire business around from wireless to non-wireless. This I did by utilizing many of the parts which we had in stock, turning them into telegraph instruments and other electrical instruments and devices. But the best remedy proved to be a large quantity of radio parts made into a sort of kit which I called "The Boy's Electric Toys."

This set, with an instruction book, showed how to conduct some 100 electrical experiments, and practically all the material that went into the kit had originally been parts used in our wireless instruments.

Before this, however, I had already sold *Modern Electrics* to a former partner of mine, and this magazine was later merged with *Popular Electricity* and became known under the name of *Worlds Advance.*

This later became what is today *Popular Science.* After that I started the magazine *Electrical Experimenter,* which also made history. Like *Modern Electrics,* the *Electrical Experimenter* was also closely connected with the E. I. Co. It was destined to become a great radio magazine, as it carried more radio text and more radio advertising than any other magazine of its time.

When the World War had gone into its second year, many Americans, myself included, began to think that it would be only a matter of time until the United States, too, would enter the war. With the passing of the Wireless Act of 1913, the usefulness of the *Wireless Association of America* had come to an end and, though up to late 1912 there had been enrolled not less than 22,300 members, the *Wireless Association* had no political significance of any kind. Late in 1915, however, I took it upon myself to form the *Radio League of America,* because I felt that the U. S. needed a body of radio recruits who would be needed when we finally did enter the war.

As the first move I called up Capt. W. H. G. Bullard of the U. S. Navy and asked him whether he would help me in forming the *Radio League of America.* This he did, early in November, 1915, by accepting honorary membership in the *League.* The board comprised the following illustrious members:

Capt. W. H. G. Bullard; Prof. Reginald Fessenden, of radio fame; Nikola Tesla, that indefatigable electrical genius who is today possibly the greatest living electrical scientist; Dr. Lee de Forest, Father of American Radio; I myself became manager. I give here a

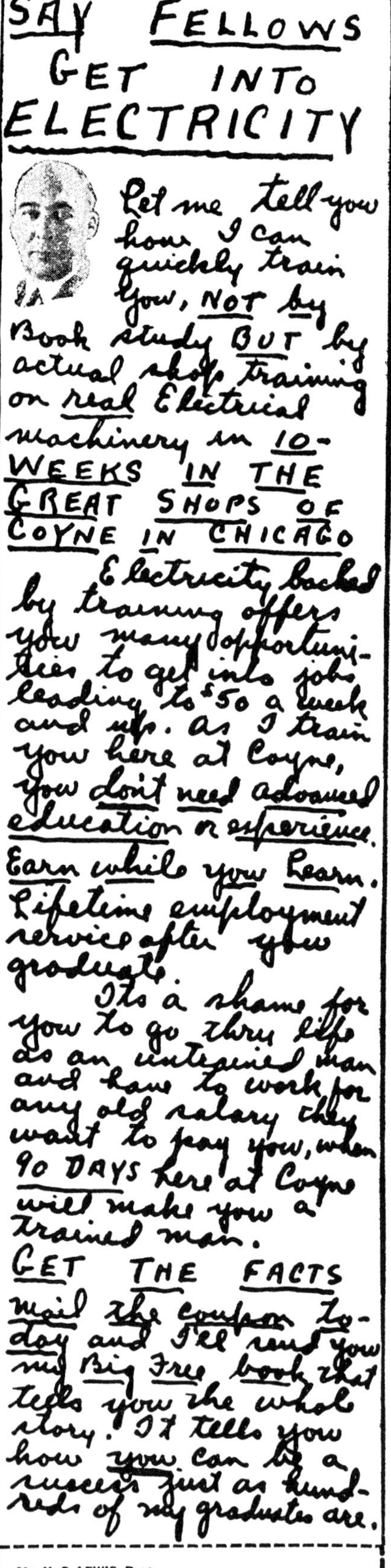

Mr. H. C. LEWIS, Pres.
COYNE ELECTRICAL SCHOOL
500 S. Paulina St., Dept. 38-83, Chicago, Ill.
Dear Mr. Lewis: Send me your big free catalog and tell me all about the **Free Electric Refrigeration** course you are including.

Name..........

Address..........

City.......... *State*..........

REPRODUCED FROM OLD ADVERTISEMENT

few excerpts from the prospectus of the *Radio League of America,* reprinted from the December, 1915, issue of the *Electrical Experimenter:*

"The advent of the great European war in 1914 found the U. S. in an unprepared condition as regards its defenses and vigorous steps were promptly taken to wake us from our lethargy. Pres. Wilson's recommendations to the country for a vast increase of our Army and Navy has been so much discussed of late that no further reference to his valuable advice is required here. It suffices to say that probably a vast majority of citizens indorse the President's defensive military program.

"But there exists today a formidable defense weapon, which up to now has not been exploited by Uncle Sam. We refer to the thousands of amateur radio stations scattered broadcast through the entire length and breadth of this fair land. There is hardly a hamlet today which does not boast of several amateur wireless stations, and their number is increasing by many hundreds each day.

"If Uncle Sam grants the amateur the free use of the ether, it is certainly up to the amateur to give something in return for the privilege. It was with this thought uppermost in his mind that Mr. Gernsback in July, 1914, first conceived the idea of organizing the *Radio League of America.*

"By referring to the 1914 Government book, 'Radio Stations of the United States,' it will be seen that only 3,723 amateurs have been licensed since 1913. The reason for this surprisingly small registration is found in the fact that the law does not require receiving stations to be licensed, nor small sending stations located in the interior of large states, where the effect of a weak spark coil would not extend over the state borders. Such stations are exceedingly numerous and have been estimated to run above 300,000. Now, then, there appears no reason for doubt that, sooner or later, the Government would pass a new law requiring the registration and licensing of such stations in order to have such stations available in case of national stress.

"No one can foretell what surprise such a new law will bring the amateurs, and for that reason it cannot be denied that it is far better and more patriotic to give this necessary information voluntarily to the Government, instead of waiting till a new law is passed which might perhaps be detrimental from the viewpoint of the amateur."

Membership in the *League* was free; there were no dues or fees of any kind. The radio amateur had merely to sign the blank on which was stated, among other things: "I the undersigned, a radio amateur, am the owner of a Wireless Station described in full on the back of this application. My station has been in use since........ and I herewith desire to apply for membership in the *Radio League of America.* I have read all the rules of the *League* and I hereby give my word of honor to abide by all the rules, and I particularly pledge my station to the U. S. Government in the event of war, if such occasion should arise.

"I understand that this blank with my signature will be sent to the U. S. Government officials at Washington who will make a record of my station."

Certificates of membership in the *League* were also made, wherein the amateur pledged himself to abide by all the rules. The *League* also issued an official insignia or button. The U. S. Government found the idea valuable and, indeed, I was in receipt of many letters from high officials endorsing the movement. Late in November, 1915, Capt. Bullard wrote to me as follows:

"I beg leave to acknowledge the receipt of your letter of November 8, 1915, submitting information concerning the *Radio League of*

(*Continued on following page*)

THE OLD E. I. Co. DAYS

(Continued from preceding page)

America, together with a copy of the Certificate of Membership.

"It seems to me that you have undertaken to carry through a very patriotic motive in banding together the great number of amateur wireless operators in the U. S. which as you state, now number probably over 300,000 of which only about 3,000 have been licensed; it being presumed that the remainder operate receiving stations only, and under the law are not required to have licenses.

"The Naval Radio Service is particularly anxious to increase its operating personnel in time of public peril when many private stations, ship and shore, would probably be taken over by the general government, and the thought has arisen that through cooperation with the *Radio League* many of its members would like to enroll themselves for active service under the Navy Department at such times as their services might be required.

"You can readily understand that any information collected by the *Radio League* will be of the greatest value to this service, and this office will be glad to avail itself of your kind offer to furnish such free of all cost, this to contain the names, locations, etc., of all amateurs in the U. S.

"I should like further to take advantage of this opportunity to ask your cooperation in enrolling members of the *Radio League* for government (Navy Dept.) operation in time of war, and take this opportunity to enclose a circular prepared in this office which we have sent out to operating wireless companies in the U. S. who in turn have distributed them to their operators, with the result that many civilian operators have engaged themselves to enlist in the Navy in time of war. Could we not do something similar to enlist members of your *League*, and would you lend your efforts to cooperate along these lines by giving the matter a little publicity? From the monthly list you propose to furnish we can get the names and addresses of many amateurs, among whom we might find many who would wish to enroll themselves. Of course, the forwarding letter accompanying the circular would have to be modified, but that could easily be done to call the attention of amateurs to how they can really serve their country in time of need.

"Please accept my thanks for thus being allowed to bring to your attention certain views in connection with the *Radio League*, an organization which can be made to be of the utmost help to the government and by focusing the attention on existing laws bring home to every amateur the desirability of cooperation all along the line to correct the great question of interference with proper government, commercial and other legitimate correspondence handled by means of the wireless art."

(Signed) "W. H. G. BULLARD,
"Captain, U. S. Navy,
"Supt. Radio Service."

From then on, the *Radio League of America* functioned smoothly and enrollments grew at a rapid rate. This was fortunate for when in April, 1917, the U. S. finally did enter the War, the radio amateurs at least were ready and prepared. I was in receipt of many communications from the government during the War seeking to recruit radio operators for the service.

Thus, for instance, on the date of June 16, 1917, Adjutant General C. J. Martin of Kansas wired me as follows:

"Radio Company, Kansas National Guard, needs ten commercial radio operators. Please furnish names and addresses such operators in this section of country. Reply my expense by night letter."

Adjutant General Martin was furnished the names promptly, and acknowledged receipt of them in his letter of June 25, 1917, as follows:

"Kindly accept my thanks for your kindness in furnishing me with the names of radio operators in this section of the country.

"Capt. Elmer J. Stahl, Topeka, Kansas, is commanding officer of the Radio Co. and it is probable that Capt. Stahl and the members of his

★ ★ ★

★ ★ ★

Hammarlund Vernier Variable Condenser

.001 Mfd. $7.00
.0005 Mfd. $6.00
3½-in. Bakelite Dial, $1.00

For Super-Accurate Tuning

The Radio News Laboratories rate this condenser 96% perfect, the highest rating given any condenser. Tested and approved by the Tribune Institute.

Write for descriptive circular.

HAMMARLUND MFG. CO., Inc.
144-146 West 18th Street
New York

THIS ADVERTISEMENT APPEARED IN 1922

organization would be interested in the electrical and radio publications published by your company. It might be well to send them a list of such publications and possibly a catalog of apparatus.

"Yours very truly,
"C. J. Martin,
"Adjutant General."

On August 15th, 1917, Lieut. C. H. McCam, of the Navy, wrote me as follows:

"The U. S. Navy Department has instructed this recruiting office to make every effort to enlist men for service in the radio branch of the Navy.

"These men are wanted immediately for active service.

"We are unable to find a complete list of all radio operators in the St. Louis district, which comprises the entire state of Missouri, with the exception of Jackson and Buchanan counties, in addition to two Illinois counties, Madison and St. Clair.

"One of your subscribers to the *Electrical Experimenter* suggested to me that you would be in a position to furnish us with a complete list of all licensed radio operators in this district, and in addition, those who are amateurs but who have not as yet been licensed.

"You may be sure that any information given us will be highly appreciated.

(Signed) "C. H. McCam,
"Lieutenant, U. S. Navy."

The lists of names were promptly supplied. About that time, I decided it was necessary to do something for those radio amateurs who had enrolled in the government service, and I started to publish their names under the caption, "*Radio Roll of Honor*." Each month in the *Electrical Experimenter* we listed the names of those who had enrolled. I continued to urge amateurs to enlist as radio operators for the government throughout 1917. I exhorted the sometimes flagging spirit of the young men who did not respond as fast as I thought they should.

At the end of the War, there being no further necessity for keeping the *Radio League of America* alive, it was disbanded.

Several years before this, however—in 1910—the E. I. Co. had moved to 233 Fulton Street after 84 West Broadway had become too small, and expansion was necessary; so the 5-story building at the famous address was rented. This address indeed became the first radio center, and in its immediate wake followed the present "*Radio Row*" adjacent to 233 Fulton St., in Greenwich and Cortlandt Sts. Of course, the great activity of Radio Row did not begin until the broadcast boom in 1921, but still 233 Fulton St. was the daddy of them all and, when the boom started, there blossomed on each side of 233 Fulton Street, as well as across the street and on the corner, one radio store after another, until the overflow ran into Greenwich and later into Cortlandt Sts.

Foreseeing the ultimate growth of radio, particularly after the war, while still on the premises of the E. I. Co. at 233 Fulton Street, I started *Radio News*, which was to become the world's greatest radio magazine in its day. This was in 1919.

During the radio boom our publishing ventures assumed huge proportions, so much in fact, that the E. I. Company became less and less important and, as thousands of others

(Continued on following page)

(Top) "MC" CONDENSER (Left) "SWK" COIL FORM

(Right) NEW "HFD" MICRO CONDENSER

Twenty-seven years ago, Hammarlund precision products were first introduced. Their exact design promptly became the talk of the radio world. Today, Hammarlund precision products are more than ever the choice of engineers, professionals, amateurs, set builders, and experimenters everywhere. Every Hammarlund product features the finest quality insulation, carefully selected metals, in addition to the famous painstaking Hammarlund design and manufacture — guaranteeing trouble-free long-lasting service.

Over 500 precision products are now being made by Hammarlund, viz.—"MC" single and split stator midget and "HF" micro variable condensers (available in 44 sizes, 20 to 320 mmf.); transmitting condensers (38 sizes—20 to 1000 mmf., 1000 to 12000 volts); acorn and standard Isolantite sockets (4 to 8 prongs); receiving and transmitting R.F. chokes (6 types, 2.1 mh to 50 mh., 60 ma to 500 ma.); "SWK" XP-53 plug-in coils (plain and wound—4, 5 and 6 prongs), etc. For dependable results use Hammarlund all the way. To be sure you get a genuine Hammarlund product, look for the name Hammarlund stamped on every product. Mail coupon for new "38" catalog with complete description of all Hammarlund products.

HAMMARLUND MFG. CO., INC. RC-3
424-438 W. 33 St., N. Y. City
( ) Please send me "38" catalog.
Name ....
Address ....
City ....State ....

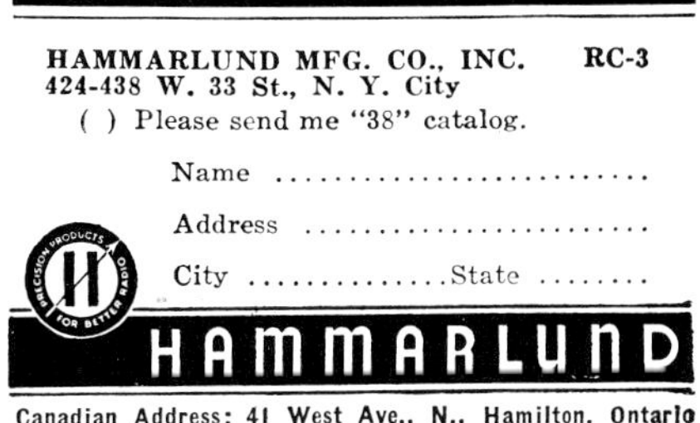

Canadian Address: 41 West Ave., N., Hamilton, Ontario

THE OLD E. I. Co. DAYS

(Continued from preceding page)

jumped into the making of radio instruments, it was thought best to concentrate our efforts on publishing. Consequently, the E. I. Company stopped its radio and catalog activities and turned publisher, and in this, the E. I. Co. was highly successful.

Between 1 and 1½ million books were sold subsequently, during the radio boom, by the E. I. Co. to Woolworth's and other outlets.

These little books sold for from 10c to 25c, and many of them are still to be found in the libraries of radio enthusiasts.

In the latter '20's the E. I. Co. book business was merged with that of the Experimenter Publishing Company, thereby closing the long and varied existence of a pioneer company that made history in its day.

IDENTIFICATION OF ILLUSTRATIONS IN THIS ARTICLE

(1) One of the first E. I. Co. catalogs in 1906.
(2) First radio set advertisement on record; advertisement appeared in the January 13, 1906, issue of *Scientific American*.
(3) Earliest adjustable zinc spark-gap for amateur use.
(4) Earliest slide-plate condenser. About 1909.
(5) Auto-coherer of 1905.
(6) E. I. Co. coherer and decoherer of 1905.
(7) First potentiometer for amateurs, by E. I. Co. About 1908.
(8) Transmitting helix (inductance). 1908.
(9) Variable transmitting condenser. 1908.
(10) E. I. Co. rheostat regulator. 1907.
(11) First E. I. Co. tuning coil with Gernsback patent ball-bearing slider. About 1907.
(12) Adjustable zinc spark gap, for amateurs. 1908.
(13) Earliest crystal detector stand. About 1908.
(14) E. I. Co. patent ball-bearing slider.
(15) Cover of E. I. Co. catalog in 1907.
(16) E. I. Co. 1-inch spark coil.
(17) Gernsback rotary variable condenser. 1911
(18) Adjustable E. I. Co. transmitting inductance.
(19) E. I. Co. bare-point electrolytic detector. 1908.
(20) E. I. Co. sealed-in electrolytic detector.
(21) E. I. Co. 1-inch spark coil.
(22) Large tuning coil with ball-bearing sliders.
(23) E. I. Co. Telimco transmitter and receiver. About 1906.
(24) First E. I. Co. electrolytic detector. About 1908.
(25) E. I. Co. adjustable transmitting inductance with pilot lamp.
(26) First E. I. Co. electrolytic interrupter. 1910.
(27) E. I. Co. transmitting key.
(28) Cover of one of the earliest issues of *Modern Electrics*.
(29) First E. I. Co. loose coupler with fixed secondary. 1910.
(30) First E. I. Co. loose coupler with variable secondary.
(31) E. I. Co. loose coupler—perfected model.
(32) Loose coupler with crystal detector and loading inductance.
(33) E. I. Co. rotary variable condenser.
(34) E. I. Co. air-cooled ½-kw. transmitting transformer.
(35) The first audion or vacuum tube sold by E. I. Co. 1911.
(36) Headquarters of E. I. Co. at 233 Fulton St., N. Y. C. 1910.
(37) First portable radio transmitter used for advertising purposes by E. I. Co. 1908.
(38) E. I. Co. rotary zinc spark gap.
(39) First complete receiving unit comprising tuning coil, telephone receiver and variable crystal detector made by E. I. Co. 1913.
(40) E. I. variable loading inductance.
(41) E. I. rotary potentiometer. About 1910.
(42) E. I. Co. complete receiving set with tuning variable condenser, detector, potentiometer, loading inductance and phones, assembled on board. 1913.
(43) E. I. Co. de luxe receiving outfit. 1915.
(44) E. I. Co. loading inductance.
(45) Circuit of Gernsback radio remote control.

Will They Have Traffic Laws In The Sky?

WILL there be policemen to tell you to "Keep To The Right" or raise a warning finger when you start exceeding the speed limit over New York or San Francisco?

Will there be "AIRPLANE GARAGES" where one may stop for gasoline or repairs?

Will the airplane be used to deliver pins or pianos? Will you step into an airplane jitney and speed down to the office in five minutes, where it now takes 30 or 35?

Who can say? Anything is possible if the airplane continues to develop as it has in the last ten years.

Right now, there are scores of positions open for men who are expert on airplane engines. And every year there will be more.

And right now is the time to get into this business—just as hundreds of men got into the automobile business and the movie business before the rush began.

Best of all, you can learn everything you need to know about airplane engines right in your own home in an hour a day of the spare time that now goes to waste.

All that we ask is this:—Fill out the coupon printed below and mail it to Scranton. This doesn't obligate you in the least—but it will bring you the full story of what the I. C. S. can do for you.

— — — — — TEAR OUT HERE — — — — —

INTERNATIONAL CORRESPONDENCE SCHOOLS
Box 5614-C, Scranton, Pa.

Without cost or obligation, please send me complete information about the subject before which I have marked an X in the list below:

☐AIRPLANE ENGINES	☐ADVERTISING
☐Mechanical Engineering	☐Salesmanship
☐Civil Engineering	☐BOOKKEEPING
☐SURVEYING	☐Civil Service
☐ARCHITECTURE	☐Railway Mail Clerk
☐Electrical Engineering	☐Traffic Manager
☐CHEMISTRY	☐Business Management
☐Stationary Engineering	☐Cert. Public Accountant
☐Automobiles	☐Stenography
☐DRAFTING	☐Good English

Name.......

Address.......

Occupation.......

Persons residing in Canada should send this coupon to the International Correspondence Schools Canadian, Limited, Montreal, Canada.

THIS ADVERTISEMENT APPEARED IN 1922

REPRODUCED FROM OLD ADVERTISEMENT

"NEW! AUDIO 'EXPRESSOR' FOR SOUND SYSTEMS"

This Public Address article, by A. C. Shaney, will appear in April Radio-Craft; due to lack of space it could not be included in this March issue as originally scheduled.

NEW TUBES INTRODUCED TO RADIO MEN —IN 1937-38

(Continued from page 629)

310A—TABLE IX

Heater Rating

Heater voltage (A.C. or D.C.)	10.0 V. (exact)
Nominal heater current	0.32-A.

Typical Characteristics

Plate voltage	135	250 V.
Screen-grid voltage	135	135 V.
Control-grid bias	−3	−3 V.
Suppressor-grid voltage	0	0
Plate Current	5.4	5.5 ma.
Output power	250	480 milliwatts
2nd-harmonic	22 to 40	26 to 30 db. below fund'l.
3rd-harmonic	30 to 55	30 to 43 db. below fund'l.

311A—TABLE X

Heater Rating

Heater voltage (A.C. or D.C.)	10.0 V. (exact)
Nominal heater current	0.64-A.

Operating Conditions

	Plate Volts	S.-G. Volts	C.-G. Volts
Recommended conditions	135	135	−15
Maximum conditions	180	150	−18

WL-706—TABLE XI

Technical Data and Ratings

Potential drop at 0.325-A. (D.C.)	109 ± 6 V.
Maximum current	600 ma.
Typical operating current	325 ma.
Minimum supply voltage	125 V., D.C.
Average tip starting voltage with respect to anode	−125 V., D.C.
Maximum tip starting voltage with respect to anode	−140 V., D.C.
Maximum starting tip current	20 ma.
Type of cathode	Cold
Gas	Argon

WL-896—TABLE XII

Technical Data and Ratings

Voltage range	5-8 V.
Current range	225-275 ma.
Max. allowable ambient temp.	150° F.

This tube must always be operated enclosed.

WL-788—TABLE XIII

Technical Data and Ratings

Voltage range	9-24 V.
Current range	225-275 ma.
Max. allowable ambient temp.	150° F.

This tube must always be operated enclosed.

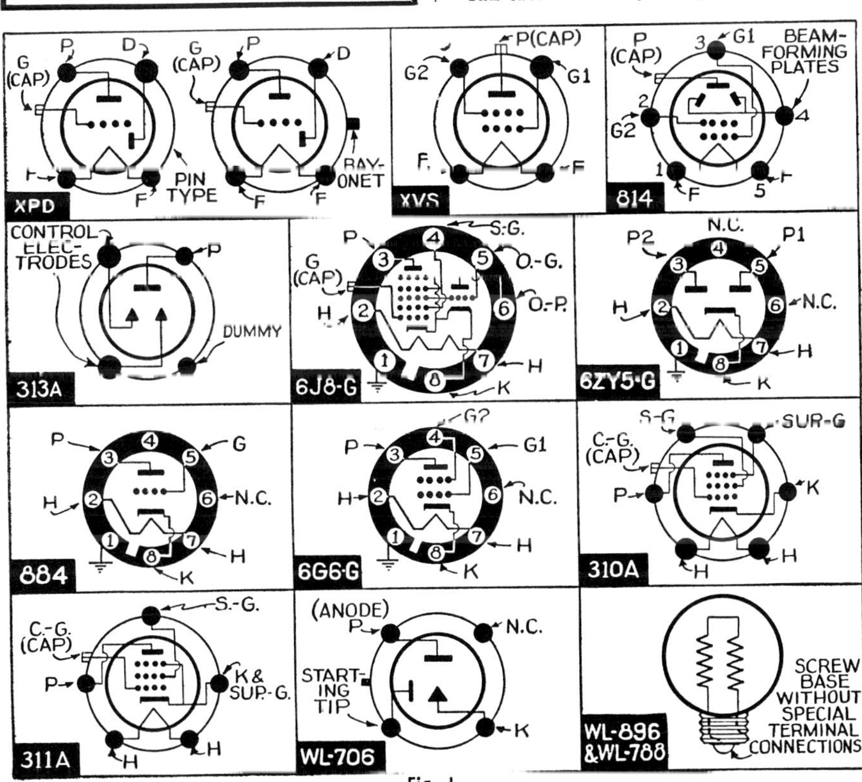

Fig. 1.
Underside view of tube sockets; "N.C." signifies "no connection".

TRY-MO *Specials for This Month!*

TRY-MO RADIO carries the largest and most complete stock of speakers, pick-ups and speaker accessories. Specialists in speaker repairs, we can repair any speaker. Write for our estimate. We can build power amplifiers and public address systems to any specifications. We can supply anything in radio. Write us for our price.

LOFTIN-WHITE AMPLIFIER

Licensed

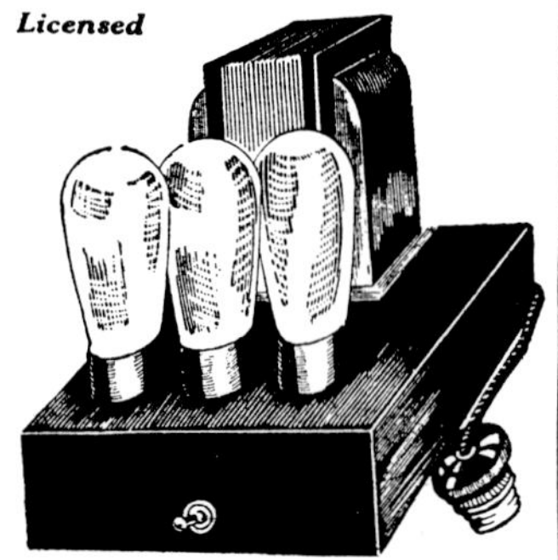

Loftin-White amplifiers in striking modernistic finish. Ideal for phonograph or public address work.

245 Model uses 1-224, 1-245, 1-280 .... **$12.50**

250 Model uses 1-224, 1-250, 1-281 .... **$15.95**

AMPLION HAND MICROPHONE

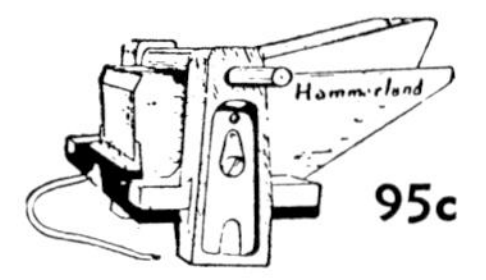

$4.95

Single button hand microphone of the tranverse current type 500 ohm type. Very sensitive and stable. A real buy.

HAMMARLUND VAR. CONDENSER

95c

Straight line frequency. .0005 mfd. maximum capacity. Brass soldered plates, low loss construction.

BEST DELUXE THEATRE PICK-UP

Unexcelled for tone quality reproduction, the BEST stands far above any other pick-up. Volume control is built in, conveniently on the base **$7.25**

BARGAIN LISTING OF LAST-MINUTE SPECIALS!

Carryola Pick up Head ..**$2.49**
Powertone 2 way Relay .. **.75**
Powertone 4 way Relay ... **.99**
Powertone Tone Control ..**$1.50**
Powertone Tone Control for panel mount **$1.05**
Air King Short Wave Coils, set of 4 ....**$1.59**

SPEAKER SPECIALISTS

TRY-MO RADIO CORPORATION
85 Cortlandt Street New York City

AMPLIFIERS Designed To Specifications

REPRODUCED FROM OLD ADVERTISEMENT

RADIO PARTS OF YESTERYEAR

(Continued from page 567)

duced-speed tuning controls are another of the early parts which have survived. They are used in practically every set today, except in the lowest price ranges.

Many parts which have existed from almost the first days of radio, have persisted until the present, often in but slightly modified form. Others have outlived their usefulness and vanished from the scene forever.

It is interesting to note, in concluding this necessarily sketchy account, that many commercial radio parts had their genesis in a passing comment, or published account of how some thrifty or imaginative radio experimenter found an inexpensive, simple solution to a problem. It is unfortunate that these unsung "heroes" of radio's swaddling-clothes stage have become lost in the shuffle; but with the passing of the years to which we dedicate this issue of *Radio-Craft* the increasingly abundant fruits of their labor bear undying witness to the ingenuity of these indefatigable pioneers.

CRL

No. 100 Filament Rheostat for Panel Mounting

No Magnetic Material Used in its Construction

This new Rheostat consists of a resistor of special non-corroding alloy inserted in a molded base of high insulating and heat resisting properties — genuine Thermoplax. Each turn of the resistor is anchored firmly in place so that there is no chance for noisy or scratchy operation. All metal parts are nickeled.

If you cannot obtain CRL Rheostats from your local dealer, send $1.00 plus 10c. for carriage.

List Price (East of Rocky Mountains) $1.00

Dealers and Manufacturers of Radio Equipment are invited to communicate with us. We are prepared to make immediate shipments.

CENTRAL RADIO LABORATORIES

303 Sixteenth Street
MILWAUKEE WISCONSIN

THIS ADVERTISEMENT APPEARED IN 1922

A CHRONOLOGICAL HISTORY OF RADIO

(Continued from page 555)

them, allowed the golden opportunity of being the first to discover wireless slip by. By 1905, he had devised a means of wireless communication from his earlier experiments, but the Marconi system was well established by that time.

1895-1900 GUGLIELMO MARCONI, Italian. Considering the inventions and research of previous years, it is with no great surprise that we determine that scientists of this era looked upon Marconi as an interloper and one of audacity. In 1895, Marconi conducted experiments with Hertzian waves, and was able to send and receive messages over a distance of a mile and a quarter. He employed the coherer invented by Branley, with Popoff's automatic tapper for decohering after a signal was received. In fact his apparatus differed very slightly from that of his predecessors when he applied for and was granted his first patent in England in 1896 for wireless telegraphy. From then on, however, Marconi made rapid strides in the advancement of the art, being successful in transmitting and receiving messages between two warships over a distance of 12 miles. In this year, Marconi was successful in enlisting the financial backing of a number of wealthy Englishmen, and formed the Wireless Telegraph and Signal Company; he was made a director of this company and placed in charge of all development work although he was then but 23 years old. In 1899, he adapted to wireless, Sir Oliver Lodge's principles of *syntony*, or tuning of circuits, perfecting it and obtaining a patent in 1900. It was a remarkable step forward in wireless transmission and reception, since it eliminated the interference of stations transmitting simultaneously, a problem of no mean proportions until that time. In 1899, Marconi was successful in covering distances up to 74 miles with his instruments, and ship and shore stations began to install his equipment. His activities and progress with wireless filtered through to America, and in 1899 he was invited to this country by the New York *Herald* which engaged him to report the international yacht races held in October of that year. Marconi accepted for another reason, he wanted to interest the United States Navy in his equipment in the hope that it would make large purchases and thus help commercially exploit wireless. To facilitate matters, representatives of the British company financed and incorporated the Wireless Telegraph Company of America, to take care of the Marconi interests in this country. Marconi then went ahead with the transmission and reception of the yacht race results, and an amazed American public obtained the news as to who had won, long before the ships had returned to port. From this angle Marconi's efforts were thoroughly successful, but not so with the Navy. In demonstrations, the official witnesses were considerably impressed by the efficiency of his equipment, although in their reports mention was made of the interference obtained when two transmitters were operating. Marconi, with the success of his experiments with Lodge's syntony or tuning still fresh in his mind, specified that this defect could be overcome. The deciding factor, however, against Marconi's equipment was the terms of his proposed contract, which the Navy definitely rejected. Thus, for a while, no further real progress was made in wireless in this country. Marconi, in the meantime, had gone back to England to continue with his experiments and make further rapid advances in the art of wireless communication. His famous trans-Atlantic transmission of the letter "S" is described elsewhere in this issue.

1900-1905—REGINALD A. FESSENDEN, and LEE DE FOREST, Americans. These two gentlemen were the outstanding American contributors to the art of wireless in its earlier days, and to each has been applied the appellation of "father of American radio".

Fessenden, while fully acquainted with Marconi's wireless equipment—having experimented with these devices—was more interested in radiotelephony. He knew that Marconi's system was adapted only to damped-wave transmission and that, as such, would not tolerate super-imposing on it voice or further irregular waves. Consequently, he began to experiment with *continuous wave* transmissions (now known as C.W.), which led to his perfecting an arc transmitter. However, the coherer would not receive the voice impulses modulated on the oscillating wave produced by the arc; so, remembering his electricity and chemistry, Fessenden created the *electrolytic detector*, which allowed current to flow in only one direction. It consisted of a small aluminum cup, filled with a solution of acid and water into which a fine silver wire dipped, which was a tremendous improvement over the coherer, and increased the receiver's efficiency considerably. Later on, Fessenden conceived the idea of employing an alternator, similar to a regular A.C. generator but with a frequency much higher than 120 cycles—to an antenna (similar to the arc transmitter's) and thus eliminating the spark gaps and arcs which wasted so much power. While at the time he was laughed at, his idea was in the future to play a very important part in the progress of radio.

Meanwhile, de Forest was experimenting with wireless, and in 1901 built an outfit less cumbersome and more efficient than Marconi's. He, too, employed the electrolytic detector, which caused between him and Fessenden considerable legal conflict which later was determined in Fessenden's favor. De Forest secured some financial backing and formed the American De Forest Wireless Telegraph Company. With this company he commenced manufacturing equipment, some of which he sold to the Army. Unfortunately, the company depended upon stock promotion for capital to finance its development work, and soon it was in financial difficulties that hampered it from getting into the commercial communications field. In this same period, 1904 to be exact, J. AMBROSE FLEMING, English, developed his 2-element (diode) "valve" while employed by Marconi. He remembered Edison's experiments and the so-called "Edison effect"—since he had been a scientific adviser to the Edison Electric Light Company of America—and hence it occurred to him that the phenomenon could be employed to advantage as a detector of radio waves. This invention was to enjoy only a short life, inasmuch as de Forest's discovery of the 3-element (triode) or audion tube was soon to follow.

1906 DE FOREST'S *Audion.* Here is the mightiest radio invention of all! It consisted only of the insertion of a grid between the filament and plate of Fleming's "valve", yet this addition of a third element so revolutionized radio that today we must be grateful for its conception. While the power or the ability of the audion tube as an amplifier or generator of oscillations had not as yet been recognized, its merit as a detector was soon proven. Despite this invention, and other meritorious work in the wireless field, de Forest's finances were in extremely poor shape. To obtain the necessary capital, he was forced to sell stock in his company, but somehow an unwilling public could not be interested. Later on, in 1912, to obtain funds for himself and his company, de Forest sold the rights to the Audion amplifier to the American Telephone and Telegraph

(Continued on following page)

A CHRONOLOGICAL HISTORY OF RADIO

(Continued from preceding page)

Co., for a paltry sum compared to its actual worth.

1907 *First Crystal Detector,* BY G. W. PICKARD. Up to this time, the most popular detector was the electrolytic type; the coherer, while still somewhat used, having been found unstable and insensitive. The Fleming valve was never really popularized, because of its insensitivity to weak impulses. Consequently, the development of the *crystal detector* marks another great stride in the development of radio. While the first employed silicon as the mineral, it was later determined that galena, iron pyrites, and many other minerals (even carborundum) also are efficient. It was extremely effective as a detector of feeble irregular impulses (modulated C.W., damped waves) although somewhat critical in the adjustment of the "catwhisker". Because of its inexpensiveness, it was the most popular of all detectors until the advent of cheap commercial audions, and was to a great extent responsible for increased activity and interest in wireless or radio (these terms are synonymous).

1909 *S. S. Republic* and JACK BINNS. By this year, practically all large ocean-going vessels had been equipped with wireless apparatus, since it served as a means of contact with land. Fortunately so, for on January 23rd of this year the White Star liner *Republic* rammed the *Florida* off Nantucket Island, and commenced immediately to sink. Jack Binns, the wireless operator on the *Republic,* broadcast his famous "CQD" (now "SOS") which brought rescue ships that saved all but 6 of the entire crew and passengers. This drama, so tense and poignant, was reported to the entire world, and created such a favorable impression on the public's mind that wireless was definitely established for ship communication.

1911 *Ships Require Wireless.* As a result of the *Republic* episode, Congress passed an act (signed June 24th, 1910) which made it unlawful for any ship, whether foreign or American, plying between United States ports at least 200 miles apart, to leave or attempt to leave these ports without wireless equipment in good working order. Such apparatus was also required on all American ships clearing for foreign ports.

1912 *Titanic Disaster.* When this great liner struck an iceberg in mid-Atlantic on its maiden voyage, its wireless calls for help (the first "SOS") were received and picked up by the *Carpathia* which managed to arrive in time

INSURE YOUR FUTURE!

Start Training Now in

PRACTICAL RADIO ENGINEERING

Do you ever sit down and "look your job in the face"? Are you satisfied with your present position, or anxious to get ahead? Today, you can't be satisfied, because rapid new developments mean Radio wants only men who are TECHNICALLY TRAINED.

Better Men Get Better Jobs

Right now is the time to start training—so you won't be left behind when others go ahead. CREI offers you home-study or residence school training that will equip you to keep pace with Radio.

WRITE TODAY FOR FREE 48-PAGE CATALOG

"Tested Plan for a Future in Radio"

An informative, illustrated booklet that answers your every question and shows actual equipment and complete outline of CREI courses.

CAPITOL RADIO ENGINEERING INSTITUTE
Dept. RC-3
Washington, D. C.—14th & Park Rd.
New York City—29 Broadway

WATCH
A. C. SHANEY IN 1938

EVERYONE knows how his perfection of sound amplification has served as a pattern for the entire industry. And now Mr. A. C. Shaney has started 1938 with an announcement of the ACA Expressor—a new unit designed to operate with all existing amplifiers. Easily adds both Automatic Volume Expansion and Automatic Audio Volume Control to all old and new amplifiers without circuit changes. Can be connected in a minute.

An entirely new series of ACA Amplifiers are now offered at lower prices than ever before. For complete details, write or wire today.

AMPLIFIER CO. OF AMERICA
37 West 20th Street, New York City, New York

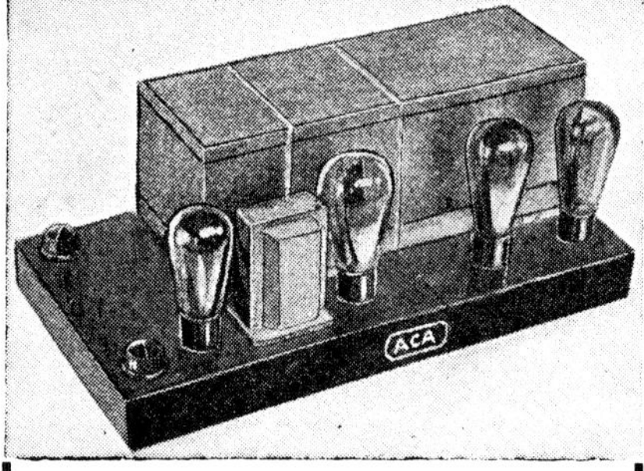

ANNOUNCING

a new engineering division headed by Mr. A. C. Shaney, an authority on amplifier design and construction. The facilities of this division will be exclusively devoted to the solution of field problems involving the use of an amplifier for sound reinforcing systems. **Send your problems to Amplifier Engineering Division for expert advice.** 16 New model amplifiers available to meet the requirements of the smallest or largest installation. Send 10c for a construction bulletin of 112, 171, 210 and 250 Power Amplifiers.

AMPLIFIER CO. of AMERICA
20 West 22nd Street New York, New York

"DEPEND ON DEPENDABLE"

DON'T BUY TEST EQUIPMENT

until you've investigated Radio City products. See your jobber. Write to Dept. C for catalog.

RADIO CITY PRODUCTS CO.
88 PARK PLACE NEW YORK CITY

ANNOUNCING

A Complete Line of

DEPENDABLE

TEST INSTRUMENTS FOR RADIO SERVICE MEN

RADIO CITY PRODUCTS COMPANY
48R WEST BROADWAY NEW YORK CITY

"HAM" SPECIAL Standard Teleplex

A highly efficient code teacher using heavy specially prepared waxed paper tape, having two rows of perforations. Write for Free folder A-3.

We are the originators of this type instrument

Instrument with tapes prepared by expert and complete course; all for $11.95. Without Oscillator.

TELEPLEX CO. 72-76 Cortlandt St., N. Y. C.

REPRODUCED FROM OLD ADVERTISEMENTS

The Royal

is the typewriter with the rapid-fire action and adjustable *personal* touch—the machine that fires letters as an automatic gun spits bullets!

•

Built for the Expert Typist-Better for the Ordinary Operator

The new Royal way of sending ***direct force*** from the fingertips to the type takes the "grind" out of typewriting and increases the daily output of any stenographer.

•

Get the Facts!

Send for the "Royal man" and ask for a DEMONSTRATION. Or write us direct for our new brochure, ***"Better Service,"*** and a beautiful Color-Photograph of the ***New Royal Master-Model 10.***

•

ROYAL TYPEWRITER CO., Inc.
2 Park Ave., Royal Typewriter Building
NEW YORK

THIS ADVERTISEMENT APPEARED IN 1915

to pick up many survivors. An unfortunate incident was brought out later, when it was discovered that one ship had been much nearer, and could have saved many more people—if it had not been that only one wireless operator was employed on that ship and that he was "off-watch" at the time. It resulted in an amendment to the Radio Act of 1910, requiring that two operators be employed on a ship, so that a constant watch could be maintained. Out of this story emerges a new figure in radio—David Sarnoff, now president of the Radio Corp. of America. At the time he was stationed at the Wanamaker Radio Station in New York City, and received the signals between the distressed ship and its rescuers, the reports concerning the rescue work and, finally, a list of the survivors, so that an anxious world could be advised of the consequences of this tragedy.

1913 ARMSTRONG *and Regeneration.* The increase in sensitivity, which results when regeneration is introduced into a receiver, is known to all who have experimented with radio circuits. Small wonder, then, that great litigation resulted between Edwin H. Armstrong, a Columbia student at that time, and de Forest as to who was the rightful inventor. Armstrong sought a patent in 1913, whereas de Forest claimed to have discovered it with an assistant (Van Etten) in 1912, although they did not bring it out at the time. De Forest produced notebooks to prove that he discovered feedback and the oscillating properties of a tube; and, after much partisanship in the radio profession, and various court decisions, he was finally awarded the patent by the Supreme Court in 1934. Also in 1913, Irving D. Langmuir, prominent physicist with the G.E. laboratories, discovered his process for creating high vacuums. (Arnold of Western Electric developed a high-vacuum process that assumed importance in telephone work.)

1914-19 ALEXANDERSON'S *Alternator—World War.* Ernst F. W. Alexanderson, Swedish-American, had helped Fessenden build some of his earlier alternators. Due to this experience, Alexanderson was able to improve this unit so that "smooth" continuous waves with a frequency of 50,000 to 100,000 cycles per second could be generated. So great were the possibilities of his device that Marconi himself came to the G.E. laboratories in Schenectady to see a demonstration of it. As a result, the British Marconi Company began negotiating for the machine, but no immediate sale was made. A stalemate in the negotiations was reached when the United States entered the World War, and seized or closed down all private wireless stations. Throughout the war wireless was of substantial aid to

(Continued on following page)

THOUSANDS EVERYWHERE ACCLAIM THIS

New Psychology Of Life!

Have you unrealized hopes? Are the better things of life always just beyond your reach?

Times have changed—but have you? Adopt a new psychology of life and MASTER YOUR PROBLEMS. It takes no greater mental effort to achieve results when you know how.

Let the Rosicrucians show you how, by the use of simple laws, you can apply the powers of your mind to bring about startling changes in your life. If you are sincere in your desire, write for the free Sealed Book. It will point out how you may obtain this most helpful information. Address: Scribe T.B.V.

The Rosicrucians
-AMORC-
SAN JOSE, CALIFORNIA.

Perpetuating the Ancient Rosicrucian Secret Teachings

That the Sun moves north as men travel south, until finally it shines continuously in the north. For 1,500 years men laughed at the "gullibility" of the ancients, until, by sailing past the Equator, Europeans of the Middle Ages DISCOVERED what the Egyptians KNEW thousands of years before.

Suppressed Knowledge of Ancient Sages

—lost to the world for 3,000 years, but preserved by the FEW who could appreciate and use it—is available to you, through association with the Rosicrucians. Their unique and successful methods of mastering life's problems and developing Personal Power are even more effective today. Try their formulas on your own personal and practical **problems.**

Successful? — Happy?

Have you realized the highest possibilities of which you are capable? Are you going through life with your greatest assets—social, business, intellectual—locked in a chest, without looking for the key?

Send for a Free Copy of "The Secret Heritage." It may open up a new world of personal satisfaction and practical achievement; it might have a startling effect upon your own future. It is sent free to the serious. Address Scribe E.Z.J.

The ROSICRUCIANS
San Jose AMORC California
(Not a religious organization)

FACTORY TO YOU

NEW REMINGTON NOISELESS PORTABLE!

10¢ A DAY

Famous Remington Noiseless Portable that speaks in a whisper, only 10c a day, direct from factory. Standard keyboard. Automatic ribbon reverse. Variable line spacer and all the conveniences of the finest portable ever built. PLUS the NOISELESS feature. Act now. Send coupon TODAY for details.

You don't RISK a Penny

We send you Remington Noiseless Portable for 10 days' free trial. If not satisfied, send it back. We pay all shipping charges.

FREE Typing Course and Carrying Case

You will receive FREE a complete simplified home course in Touch Typing. Also FREE, a handsome, sturdy carrying case. No obligation. Mail coupon for full details—NOW

Remington Rand Inc., Dept. 189-3
465 Washington St., Buffalo, N. Y.

Tell me, without obligation, how to get a new Remington Noiseless Portable, plus Free Typing Course and Carrying Case, for 10c a day. Send Catalogue.

Name ....
Address ....
City .... State ....

Since its foundation, it has been the policy of this Company to embody in the

Remington

in perfected form, the best typewriter ideas by whomsoever advanced.

For our latest manifestation of this policy, inspect the new

Visible Writing Remingtons

which embody every desirable feature extant—PLUS an Adding and Subtracting Mechanism, which constitutes an innovation.

The voice that cried in the wilderness 30 years ago: "You cannot afford to write in the old way," now acclaims with equal conviction: "You cannot afford to calculate in the old way."

Remington Typewriter Company, Incorp.
325-327 Broadway, New York City

Vest Pocket Adding Machine

Sturdy Steel Construction. Lifetime Pocket Companion. Adds, Subtracts, Multiplies, Divides. Capacity, 999,999.99. A real machine—guaranteed 5 years. Send name and address. We ship immediately. On delivery, pay postman $2.50 (plus few cents postage). (Foreign countries send cash.) Your money back after 10 days trial if not satisfied. Thousands of satisfied users. Send for Special Offer to agents.

Free Trial Send no money $2.50 Total Cost

TAVELLA SALES CO., 21 West Broadway, New York, N. Y.

Mascot Slide Rule

It fits the pocket and your pocket book. Will quickly solve any problem in multiplication, division, proportion. Gives the square, the square root, logarithm and reciprocals of all numbers. Trig scales give sines and tangents of numbers. Fine black graduations on white celluloid. Price in fabricoid case $1.00, in leather case $1.25. Money back if not satisfied. Special offer to agents on both items. G. Tavella Sales Co., 25 West Broadway, New York, N. Y.

RADIO COURSES

RADIO OPERATING: Prepare for Gov't License Exam. ● RADIO SERVICING: Including Short Wave ● AMATEUR CODE ● ELECTRONICS ● TELEVISION ●

Day and Evening Classes—Booklet Upon Request

New York YMCA Schools
4 W. 64th Street, New York City

Wireless Instruction

Our graduates now with the Marconi Wireless Telegraph Companies, United Fruit Co., National Electric Signalling Co., Kilbourne and Clark, De Forest Co., Telefunken, Private Yachts and many others.

Send for Catalogue

Y. M. C. A. Radio School
145 E. 86th St. (*Founded 1910*) New York, N. Y.

REPRODUCED FROM OLD ADVERTISEMENTS

A CHRONOLOGICAL HISTORY OF RADIO

(Continued from preceding page)

both sides as a means of constant contact and communication and as an aid to espionage. In the United States, the Bell System (A.T.&T. Co.) was hard at work perfecting the vacuum tube which it purchased from de Forest. Its ability to function as an "oscillator", or generator of high frequencies, was established by that time, by virtue of de Forest's and Armstrong's feedback circuits. A means for modulating voice impulses on the carrier wave which was produced, also by using vacuum tubes, was developed in 1914-'15 in the G.E. labs. by Alexanderson and by Colpitts in the W. E. labs. The Hartley (W. E. Co.) oscillator circuit was developed in 1915.

These inventions resulted in experiments in radio telephony, for the purpose of facilitating and improving long-distance transmission of speech. The first test made by Bell Telephone engineers was in 1915. A low-powered transmitter was installed at Montauk, L. I., and an amplifier-receiver at Wilmington, Del. Wavelengths of from 800 to 1,800 meters were employed, and the results obtained were satisfactory enough to warrant further tests with higher power.

While the initial tests were made with transmitting tubes which totaled a maximum of 15 watts power output, before the end of 1915 several hundred such tubes in parallel (sometimes as many as 500) were employed to achieve higher power. Larger transmitting tubes, of the order of 100 and 500 watts and 1 kw., were not to be developed until some time afterwards. As a result of all these researches, in 1915 the first trans-Atlantic (and, accidentally, trans-Pacific to Honolulu) radio telephone conversations were successfully held between Arlington, Va., and the Eiffel Tower in Paris, France. The U. S. Navy, W. E. Co., and A. T. & T. Co. collaborated.

All of this was the forerunner of broadcasting which, commercially, didn't make its appearance until 1920-'1. The developments in speech transmission without wires made in this period were to form the nucleus of the equipment for the broadcasting station which was soon to come.

In 1919, Frank Conrad of Pittsburgh, Pa., an amateur and Westinghouse engineer, began broadcasting record programs from his amateur radiophone station located in a garage at the rear of his house. They were received with such great enthusiasm by other amateurs in the vicinity, who incidentally invited their friends and neighbors over to hear the "wireless music", that in a short time much newspaper publicity was given to his broadcasts. As a result Westinghouse officials, in 1920, decided to build a large station to conduct broadcasting

THIS ADVERTISEMENT APPEARED IN 1924

REPRODUCED FROM OLD ADVERTISEMENT

for the publicity and prestige that it would give the company. The station was rushed and launched in time for the broadcasting of the Harding-Cox presidential election returns—with Frank Conrad's little station standing-by just in case of an emergency. The large station later on became KDKA, now known to practically everybody with a radio receiving set. From one station in 1920, to 400 in 1922 and over 1,400 stations by 1924 was the record set by broadcasting. Further details concerning this phase of the radio industry are given elsewhere in this issue.

1919-1921 *Formation of the Radio Corp. of America.* Heretofore, the British Marconi Company had dominated in all activities of the wireless field. Their early start and strong finances permitted them to buy in and control all major patents and activities so that even here in America, their dominance was felt. After the World War, the Marconi Company resumed its negotiations for the Alexanderson alternator. At this stage, the United States government intervened since it was felt, at that time, that the sale of this American equipment might result in world domination of wireless communication by foreign interests. After conference with the Navy, another meeting was held at the General Electric offices, when it was decided to retain the alternator in the interests of this country. It was probably at this latter conference that the plans were formulated for a strictly American-owned radio company, since we find such men as Admiral Bullard, Owen D. Young and C. W. Stone of General Electric attending it. At any rate, on October 17th, 1919, the Radio Corp. of America was organized and a patent pool of heretofore competing patent interests was effected. On November 20th of that year, the assets and business of the Marconi Wireless Telegraph Company of America were taken over by RCA, and from then on American wireless was "on its own."

The story of broadcasting, how and when it first began, is told in other pages of this issue. So, too, are the stories of the advance of the vacuum tube, circuits, parts and receivers, and so-on. Wherever possible, dates are included—consequently, to repeat such information in this chronicle would be superfluous.

BIBLIOGRAPHY

"The Hysterical Background of Radio"—R. P. Clarkson, 1927, J. H. Sears and Co., Inc.
"The Electric Word—The Rise of Radio"—Paul Shubert, 1929, The Macmillan Company.
"Old Wires and New Waves"—Alvin F. Harlow, 1936, D. Appleton-Century Co., Inc.
Bell Telephone Quarterly, Vol. XVI, No. 4 (Oct. 1937).
Western Electric Co. historical compilation—special to *Radio-Craft.*
"Milestones of Radio Progress"—pamphlet, Radio Corp. of America.
"The History of Broadcasting in the United States"—pamphlet, Westinghouse Elec. & Mfg. Co.

THE RADIO MONTH IN REVIEW

(Continued from page 597)

for photoelectric scanning, but no record at receiver. *Radio-Craft* **understands that such "commercial television" does not come under F.C.C. control so long as it remains** ***intra*-state.**

Two large-screen television demonstrations in London, last month, of the Baird and Scophony systems, indicated that theatres will introduce feature, provided postmaster general, who also controls telegraph and telephone lines, permits. The Baird system showed pictures in color on 6 x 8 foot screen; but sharpness is lost as compared with black and white.

RADIOCOPS HEARD ACROSS OCEAN

WHEN Mount Vernon, N. Y., headquarters W2XNR called 3 policemen to extinguish a brush fire one morning last month, it was not exactly private. Shortly arrived a request verification from George Garvey, Liverpool, England, fan—who heard it on a 2-tuber and an 8-foot indoor aerial. And other police cars, here and there, are having problems when messages are picked up, *sometimes from the next state!*

The Havana radio conference last month voted in favor of a radio police network to cover the American continent. Recently Commissioner Valentine, of New York, sent a test message to Chief Martinez, of Mexico City, over an all-police communication hookup, which it is proposed in the near future to make routine.

BROADCASTING RADIODDITIES

RECORD chimes, used with a P.A. system for sales promotion, finished the holiday task at Plainfield, N. J., last month when they joined with the city's church bells in a Christmas morning chorus. (See photo.) Chimes were also thus broadcast 3 times *daily* during Xmas.

(Photo—R. H. Sears)

Yule carols blanketed Plainfield, N. J., via the four 6-ft. loudspeakers here shown being installed (for 4th year) by Verne M. Wintermuth (left) and James P. McClary. The Chamber of Commerce and First National Bank collaborated.

In a close-of-season game between Richmond, and William and Mary U.s, last month, Station WRVA broadcast the huddle of the Richmond team at the opening of the final quarter. But W. & M., not having sets and phones, got no tip-off from the proceeding. Forewarned is forearmed; an encore might be risky.

The mike was crashed at WHBL, Sheboygan (Wis.), by an outsider last month. A great snow-owl flew into the power wires, and put the station off the air for 35 minutes. (He is off the air permanently, himself.)

Standard R.F. transmissions have been known for years, and occasionally standard A.F. pitches. C.B.S. last month announced that, from now on, time signals will be the standard international A (440 cycles) for the use of musicians.

Debate by phonograph record will be carried on between the College of the Pacific and the University of Redlands, says an announcement received last month. Both institutions have Universal recording machines; and first the pro and con, and then the rebuttal and surrebuttal will be "disced" for radio use.

While educators last month were panning radio programs at Chicago, Radio Supervisor Blanche Young of Indianapolis suggested that important current-event broadcasts, as well as air historical pageants, be canned and made available for classroom reproduction.

RADIODDS AND ENDS

ARE there "sidebands"? Some light may be thrown on that bitterly-argued question by experiments announced last month by Prof. R. R. Ramsey, well-known radio textician. Reconciling previously conflicting experiments, he showed that, when 2 distinct notes are sounded, the ear hears a "beat-note"; but a resonator does not receive it. If a single source of sound gives off 2 notes at once, on the other hand, the beat-note can be physically demonstrated. The conclusion is that the ear creates "beat-notes," and supplies missing sum and difference-frequencies.

The French radio industry celebrated *Une Grande Quinzaine* (15-year anniversary) of broadcasting last month with the slogan: *"Un foyer sans T.S.F. est un foyer sans joie."* (A home without radio is a home without happiness.)

The American Medical Association, setting up standards for ear-testing apparatus (audiometers) last month ruled that they shall have a range from 128 to 8,192 cycles, or higher, with a frequency tolerance of 5%; be calibrated in decibels with steps of 5 or less; have pure tone (harmonics down at least 40 db.); and that manufacturers should assure servicing. But a test chart must wait on more definite consensus of medical opinion for approved standardization.

(Continued on following page)

THE RADIO MONTH IN REVIEW

(Continued from preceding page)

New York radio dealers last month issued an appeal for the organization of a national association of retailers to help in settling vexatious problems, including that of trade-in and resales.

"Videotron" for cathode-ray tubes is now claimed as a trade name, accurately descriptive, for "video" or sight work, by National Union Radio Tube Co.

A test of listener reaction to showmanship was made last month on a well-known Sunday evening program—or, to be exact, of showwomanship. The reaction was one of the strongest recorded in the history of radio. One trade commentator observed that "there was nothing objectionable in the mummery unless the listener supplied it by exerting his own imagination"—which said imagination brings us forward to the approaching day of television, and the question of how a "script" can be submitted in advance for station or official censorship?

Ground was broken, last month, for the erection of the A.R.R.L. station W1AW, designed as a memorial to the league's founder, the late Hiram Percy Maxim, at West Hartford, Conn. When it is completed, it will have 5 transmitters, each 1 kw., for voice and code.

Ten years ago a benevolent New Yorker fitted radio dials with Braille numbers for blind friends; 2 years ago, *Radio-Craft* described the construction of such a set, but more modern. Last month, G.E. brought out a commercial-type set with this feature; and the first was presented through Helen Keller to the American Foundation for the Blind, New York! It has keys, with raised numerals, for touch tuning.

More of the early history of short waves will, perhaps, find its way to light during the trial of a lawsuit, begun last month, by the estate of the late R. A. Fessenden, S.-W. pioneer, against RCA on patent claims whose validity the defendant contests on the basis of "prior knowledge" filed.

N.B.C. announced, last month, that it closed the year 1937 with 143 stations in its network, a net gain of 38. In power, the addition is less striking—112,900 watts or 6.2% (daytime).

The tuning pulley belt is the weakest

(Continued on page 654)

Here's 8-year-old Lillian Switzer trying out the new General Electric "Touch Tuning" radio set, specially equipped with Braille (raised) symbols, recently presented by Miss Helen Keller (right) to the American Foundation for the Blind.

ORIGIN AND GROWTH OF RADIO SERVICING

(Continued from page 551)

were in this field, Service Men were few, although some of the constructors began to dabble in it. Today, we find too many Service Men in the profession; and, too, too many constructors dabbling with servicing.

INTRODUCING TUBE TESTERS

About this time we also find testing equipment being developed and manufactured, especially tube testers. Previously, if a dealer sold a tube all he would do for the consumer was connect a battery across the filament prongs and show him that the tube would light. Since the construction of earlier tubes was such that the filament seldom outlasted its emitting properties, this method of proving to the customer that a tube was good was considered satisfactory until tubes were made better and then the fact that its lighting had relatively slight bearing on its efficiency was well established. It is entirely possible that the need for tube-testing equipment was another factor which made it necessary to obtain a Service Man-technician.

From this period on, and in step with the ever increasing technical advance in receiver design, the servicing profession or industry grew. To some extent this phase of the industry was encouraged by the manufacturers themselves, many of them going so far as to have a special department staffed with factory-trained men who could travel over the country teaching and instructing the jobber's and dealer's service personnel the art of servicing their particular models. But for no altruistic reason did the manufacturer do this. An insatiable demand for sets on the part of the public made it necessary for them to rush these sets out, often poorly produced and with insufficient engineering behind them. In the field, transformers would open, poorly-soldered joints would develop noise, sets became inoperative due to mistakes in the wiring (proving that some were never even air-tested), and other typical factory boners developed.

A.C.-SET SERVICING

By the early part of 1926 almost every Tom, Dick and Harry could fix a radio receiver. Even when battery sets began to be "electrified", salesmen and tinkerers could replace "B"-eliminator rectifier tubes or the copper-oxide rectifier in the "A" supply, or even make the installation when these units were sold to replace the batteries. Then along came the all-electric set, using type 226 tubes in the R.F. and A.F. stages, a 227 in the detector stage, and a 71A in the power stage, with a power-pack containing an 80-type tube rectifier, power transformer, chokes, filtering condensers, voltage divider system for power distribution to the various stages,—when the tinkerer began to "fold his tent and silently steal away". So many different types of tubes, not to mention the other compli-

(Continued on page 654)

and Complete Directory of all Commercial Wiring Diagrams of Receivers

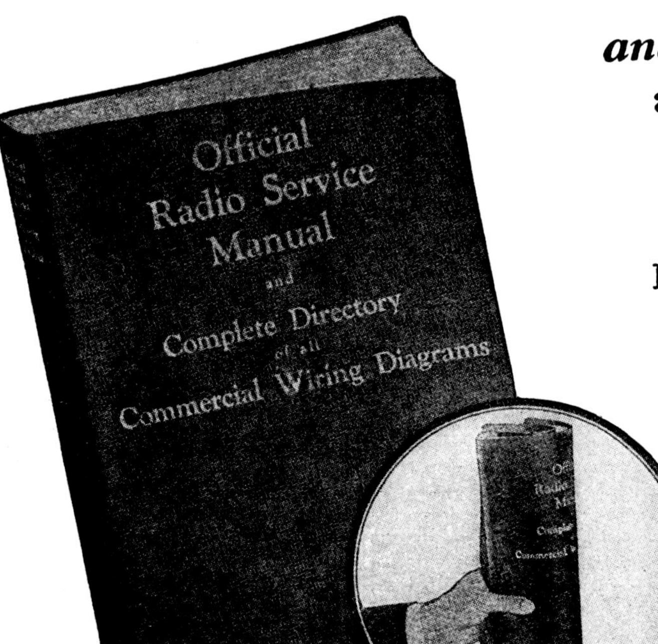

PREPARED ESPECIALLY FOR THE RADIO SERVICE MAN!

HUGO GERNSBACK
Editor

CLYDE FITCH
Managing Editor

$2.50 *Special Pre-Publication Offer*

WE ANNOUNCE the early publication of the long-awaited OFFICIAL RADIO SERVICE MANUAL. This manual is the first complete and most up-to-date book of this kind. Nothing else like it has ever appeared in print.

A tremendous amount of material has been collected, not only for the Service Man, but for everyone interested in radio. A complete directory of every radio circuit of commercial receivers is now possible, and not only do you get every circuit of every set manufactured, of which there is any record, but in addition, an entirely new idea makes it possible to keep the manual up-to-date.

The OFFICIAL RADIO SERVICE MANUAL is made in loose-leaf form—handsomely made of flexible leatherette—the entire book can be folded and slipped easily into your pocket or put in your bag.

Rarely do manufacturers supply information about receivers made before 1927—even 1930 service data are not always available because many manufacturers do not supply independent Service Men with such data. And, when you can get the material from some of the manufacturers, it is of little use to you because it is not uniform, and it is scattered in different places; difficult to get at.

Additional service data, for new receivers as they appear on the market will be published and supplied at trifling cost so that the MANUAL may be kept up-to-date at all times.

SERVICE INFORMATION

But that is not all. The OFFICIAL RADIO SERVICE MANUAL contains also a most comprehensive instruction course for radio Service Men, giving practical information from every angle on how to service sets.

In short, the OFFICIAL RADIO SERVICE MANUAL is the biggest thing of its kind that ever came along in radio. It will be hailed by every wide-awake radio man throughout the entire industry.

The size of the OFFICIAL RADIO SERVICE MANUAL is 9 in. by 12 in. There are several hundred pages, printed on good paper, and the book contains hundreds of illustrations and diagrams.

SPECIAL OFFER

The price will be $3.50 upon publication of the book; but to our own subscribers and readers, we have a special pre-publication price of $2.50, prepaid. This price will be withdrawn the minute the book is published, which will be during the next six weeks or sooner.

GERNSBACK PUBLICATIONS, INC.,
96-98 Park Place, New York, N. Y.

As per your special offer, I enclose herewith $2.50 for which you are to send me postpaid, one copy of the OFFICIAL RADIO SERVICE MANUAL as soon as it is published, at the pre-publication price of $2.50. I understand that the price will be $3.50 as soon as the MANUAL is published.

Name ....

Address ....

City .... State ....

REPRODUCED FROM AN OLD ADVERTISEMENT

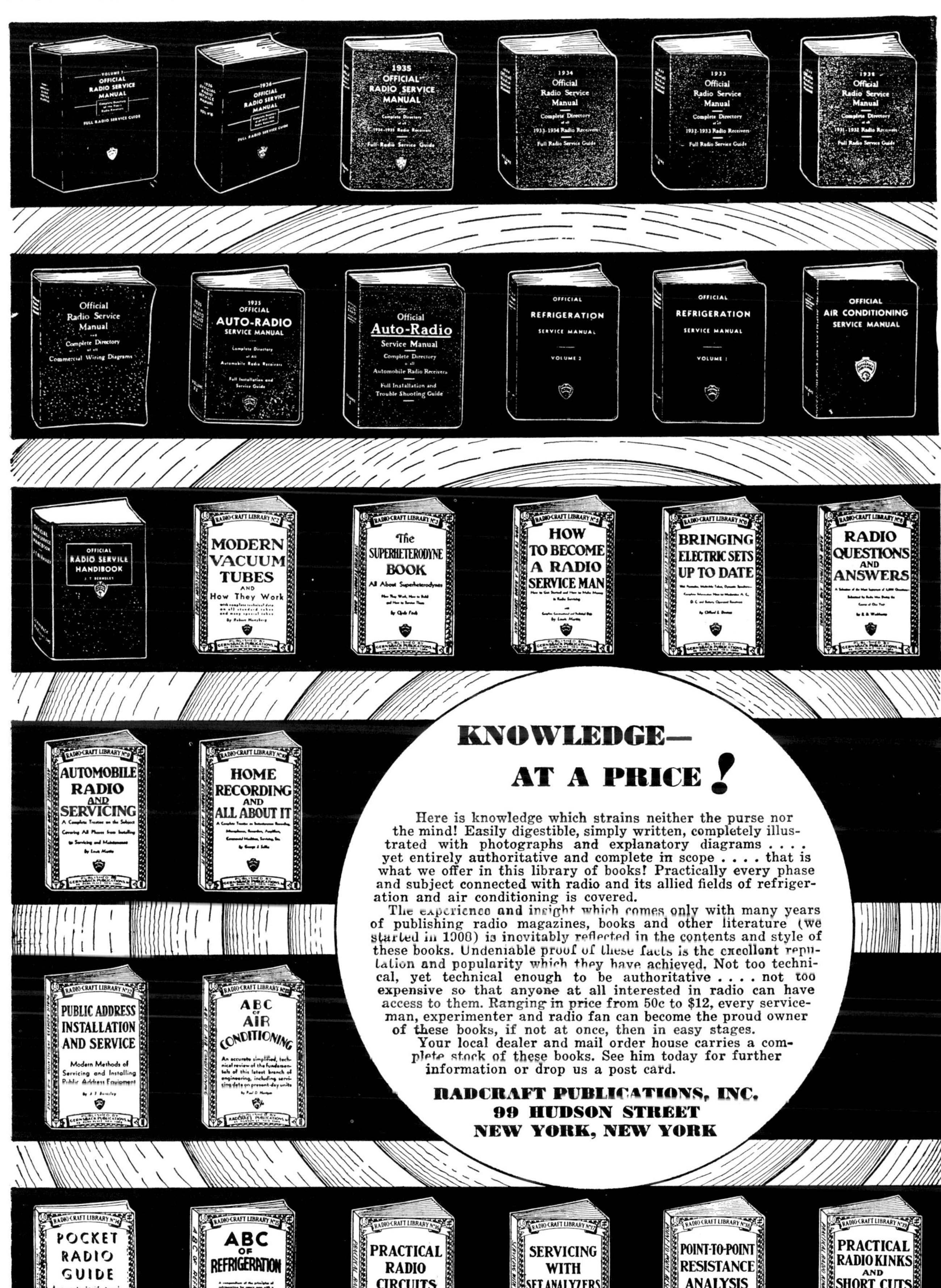
KNOWLEDGE—
AT A PRICE!
Here is knowledge which strains neither the purse nor the mind! Easily digestible, simply written, completely illustrated with photographs and explanatory diagrams . . . . yet entirely authoritative and complete in scope . . . . that is what we offer in this library of books! Practically every phase and subject connected with radio and its allied fields of refrigeration and air conditioning is covered.
The experience and insight which comes only with many years of publishing radio magazines, books and other literature (we started in 1908) is inevitably reflected in the contents and style of these books. Undeniable proof of these facts is the excellent reputation and popularity which they have achieved. Not too technical, yet technical enough to be authoritative . . . . not too expensive so that anyone at all interested in radio can have access to them. Ranging in price from 50c to $12, every serviceman, experimenter and radio fan can become the proud owner of these books, if not at once, then in easy stages.
Your local dealer and mail order house carries a complete stock of these books. See him today for further information or drop us a post card.
RADCRAFT PUBLICATIONS, INC.
99 HUDSON STREET
NEW YORK, NEW YORK

ORIGIN AND GROWTH OF RADIO SERVICING

(Continued from page 651)

cations, were a bit too confusing for him. Besides, the need for test equipment was more obviously necessary now; and that investment as compared to an ordinary pocket voltmeter and screwdriver was too much for one who just dabbled in repairing.

In the years that followed, receivers became increasingly complicated and each year would cause new expenditures in instruments, manuals (since circuits became too varied and involved for the average Service Man), and for text books, so that new engineering developments could be better understood. And each year would see some of the old-timers, who couldn't keep pace with the changing technique of servicing, drop out only to be replaced by new men—many of whom were not even as efficient. To this day that situation exists.

Perhaps, as in 1926, the even more complicated receiver to come (in the very near future) will cause a further thinning of the now overswelling ranks of Service Men, and be another salvation for the profession.

THE RADIO MONTH IN REVIEW

(Continued from page 650)

part of a radio set: many Service Men had suspected as much, but last month President Hoffman of Midwest Radio Corp. said it had been proved by a wear-testing machine. (This is the kind of robot that tells your radio's fortune, by giving it continuous action till something breaks. In 1 day, it will make a 4-year forecast.) Anyway, a 10-year cable was made of long-staple cotton with silk threads; while even a wire cable will go phutt in 2.

General Electric received authority last month to go ahead with an S.-W. broadcast station for South American and Far East transmissions, at Belmont, Calif. It will be completed this year, and work on 9,530 and 15,330 kc. Directional antennas will concentrate the beam, and give a better aerial path to Asia than that from Schenectady, which passes over the magnetic and geographical poles, and suffers much from atmospheric variations.

ADVERTISERS' INDEX

A

Aerovox Corporation 619
Allied Radio Corporation 603
Amperite Corporation 638
Amplifier Company of America 644
Arrow Sales Company 648

B

The Brush Development Company 650
Burstein-Applebee Company 650

C

Capitol Radio Engineering Inst. 644
Central Radio Laboratories 643
Clarostat Mfg. Company, Inc. 647
Cornell-Dubilier Corporation 525
Cornish Wire Company 651
Coyne Electrical School 531, 636
Crosley Radio Corporation 537

D

Dodge's Institute 648

G

G&G Genuine Majestic Refrigerator and Radio Parts Service 650
Goldentone Radio Company 650

H

Hammarlund Mfg. Company 639
Hygrade Sylvania Corporation 535

I

International Correspondence Schools 640

L

Lancaster, Allwine & Rommel 650
LeJay Mfg. Company 650
Lincoln Engineering School 650

M

Modell 633

N

National Company, Inc. .... Inside Back Cover
National Radio Institute 529
National Schools 647
National Union Radio Corp. 629
New York Y.M.C.A. Schools 646

O

Clarence A. O'Brien & Hyman Berman 637
Oxford-Tartak Radio Corp. 627

Q

Quam-Nichols Company 631

R

RCA Manufacturing Co., Inc., Inside Front Cover, Back Cover, 513, 515, 517, 519, 521
RCA Parts 641
RCA Radio Tubes 617
Radio Circular Company 648
Radio City Products Company 644
Radio Constructors Labs. 634
Radio & Technical Publ. Co. 609
Radio Lab. Service 640
Radio Training Association of America 638
Radolek Company 655
Raytheon Production Corp. 607
Readrite Meter Works 619
Reliable Radio Co. 651
Remington Rand, Inc. 646
John F. Rider 635
The Rosicrucian Order 646
Royal Typewriter Co., Inc. 645

S

McMurdo Silver Radio Corp. 523
Solar Manufacturing Company 654
Sprayberry Academy of Radio 623
Sprague Products Company 613
Standard Transformer Corp. 648
Superior Instrument Company 605
Supreme Instruments Corp. 533

T

Tavella Sales Company 646
Teleplex Company 646
William A. Thomas Company 650
Thordarson Elec. Mfg. Co. 611
Triplett Elect. Instrument Co. 641
Try-Mo Radio Company, Inc. 643

U

Universal Microphone Co., Ltd. 654
Utah Radio Products Company 621

W

Montgomery Ward & Company 539
Webster-Chicago 615
Weston Elect. Instrument Corp. 527
Wholesale Radio Service Co., Inc. 625
John Wiley & Sons 649
Wright-DeCoster, Inc. 649

Z

Zephyr Radio Company 648

(While every precaution is taken to insure accuracy, we cannot guarantee against the possibility of an occasional change or omission in the preparation of this index.)

ILLUSTRATION CREDITS

A.C.—Audak Company
A.K.M.C.—Atwater Kent Mfg. Company
A.R.C.—Allied Radio Corp.
B.T.L.—Bell Telephone Laboratories
C—Cinaudagraph Corp.
C.D.—Cornell-Dubilier Corp.
C.R.C.—Crosley Radio Corp.
D—Allen B. DuMont
D.C.—The Daven Company
DeF—Lee de Forest Laboratories
"G.C."—George Clark, RCA Historian
G.E.—General Electric Company
G.I.—General Instrument Corp.
G.I.C.—General Instrument Corp.
H.—Andrew Halbran
H.R.D.—Hansco Radio Devices
J.T.R.—J. Thos. Rhamstine
"K & H"—Andrew Halbran
LF—Lee de Forest
L.F.—Leo Freed
L.G.P.—Louis Gerard Pacent
M—Robert H. Marriott
M.C.G.B.—Marconi Company, Great Britain
"MCH"—WHK-WCLE, Cleveland
McM.S.—McMurdo Silver
M.E.—"Modern Electrics"
"M.H."—WHK-WCLE, Cleveland
M.V.—Magnavox Company, Ltd.
M.W.—Marconi's Wireless Telegraph Co., Ltd.
NBC—National Broadcasting Co., Inc.
N.S.A.—Noblitt-Sparks Industries, Inc.
O.T.M.—O. T. McIlvaine, Electronic Products Company
P—Pacent Engineering Corp.
"PGR"—"Popular Guide to Radio"
P.R.C.—Philco Radio and Television Corp.
P.R.&T.—Philco Radio and Television Corp.
RCA—RCA Manufacturing Company, Inc.
R.D.W.—R. D. Washburne
"RFA"—"Radio For All"
R.M.W.—Readrite Meter Works
S.B.—Shure Brothers Company
S.C.—Stromberg-Carlson Telephone Mfg. Co.
S.I.C.—Supreme Instruments Corp.
S.P.S.—Shrage Photo Service
T.—Transmitter Equipment Mfg. Company
T.A.E.—Thomas A. Edison, Inc.
T.C.—Transducer Corp.
T.E.—Thordarson Electric Mfg. Company
W.E.—Westinghouse Electric & Mfg. Co. (pages 543, 544, 545, 548)
W.E.&M.—Westinghouse Electric & Mfg. Co.
W.E.—Western Electric Co. (pages 568, 569)
W.L.—Ward Leonard Electric Co.
W.M.C.—Workrite Manufacturing Company
W.R.S.—Wholesale Radio Service Company
W.T.C.—Western Television Corp.
X.L.R.—X-L Radio Laboratories

CREDIT CORRECTIONS

Photo	Pg.	Present Credit	Correct Credit
CRL regen. set	548	none	McM.S.
W.E. horn speaker	569	B.T.L.	L.G.P.
Cascade audio ampl.	571	M.V.	L.G.P.
RCA WD11 tube	582	L.G.P.	G.C.

SPECIAL CREDITS

Radio-Craft takes this opportunity to express appreciation to several individuals and organizations whose exceptional cooperation helped make possible this JUBILEE SOUVENIR NUMBER.

Mr. George H. Clark, RCA historian, made available to *Radio-Craft* many photographs in his personal file of photos of old-time radio equipment. Mr. M. Raber, who collaborated with Mr. Clark in furnishing these photos, also contributed numerous photos from the historical archives of the Radio Corp. of America. Mr. L. G. Pacent made available for photographing by *Radio-Craft* actual equipment in his private museum of old-time radio apparatus. Mr. R. H. Marriott supplied many of the exceptionally old views, some of which date prior to 1900. Mr. Joseph D'Agostino, N.B.C. engineer, contributed generously of data and photographs of tubes in his internationally famous collection (on display in Radio City). Mr. E. J. Quinby supplied considerable material (data and photos) from the historical files of the Western Electric Co. Bell Telephone Labs. furnished photographs of equipment in their museum. Westinghouse Elec. & Mfg. Co. contributed many of the photos and much of the information concerning the early days of broadcasting.

Space limitations preclude giving full credit to the many hundreds of individuals and organizations which so wholeheartedly cooperated in making the JUBILEE SOUVENIR NUMBER a success.

THIS ADVERTISEMENT APPEARED IN 1924